D1757976

WITHDRAWN
FROM
UNIVERSITIES
AT
MEDWAY
LIBRARY

DRILL HALL LIBRARY
MEDWAY

ENG
1300
3592

SIMPLIFIED LRFD BRIDGE DESIGN

4682549

SIMPLIFIED LRFD BRIDGE DESIGN

Jai B. Kim
Robert H. Kim
Jonathan R. Eberle

With Eric J. Weaver and Dave M. Mante

CRC Press is an imprint of the
Taylor & Francis Group, an **informa** business

CRC Press
Taylor & Francis Group
6000 Broken Sound Parkway NW, Suite 300
Boca Raton, FL 33487-2742

© 2013 by Taylor & Francis Group, LLC
CRC Press is an imprint of Taylor & Francis Group, an Informa business

No claim to original U.S. Government works

Printed on acid-free paper
Version Date: 20130125

International Standard Book Number-13: 978-1-4665-6651-4 (Hardback)

This book contains information obtained from authentic and highly regarded sources. Reasonable efforts have been made to publish reliable data and information, but the author and publisher cannot assume responsibility for the validity of all materials or the consequences of their use. The authors and publishers have attempted to trace the copyright holders of all material reproduced in this publication and apologize to copyright holders if permission to publish in this form has not been obtained. If any copyright material has not been acknowledged please write and let us know so we may rectify in any future reprint.

Except as permitted under U.S. Copyright Law, no part of this book may be reprinted, reproduced, transmitted, or utilized in any form by any electronic, mechanical, or other means, now known or hereafter invented, including photocopying, microfilming, and recording, or in any information storage or retrieval system, without written permission from the publishers.

For permission to photocopy or use material electronically from this work, please access www.copyright.com (http://www.copyright.com/) or contact the Copyright Clearance Center, Inc. (CCC), 222 Rosewood Drive, Danvers, MA 01923, 978-750-8400. CCC is a not-for-profit organization that provides licenses and registration for a variety of users. For organizations that have been granted a photocopy license by the CCC, a separate system of payment has been arranged.

Trademark Notice: Product or corporate names may be trademarks or registered trademarks, and are used only for identification and explanation without intent to infringe.

Library of Congress Cataloging-in-Publication Data

Simplified LRFD bridge design / editors, Jai B. Kim, Robert H. Kim, Jonathan Eberle.
pages cm
Includes bibliographical references and index.
ISBN 978-1-4665-6651-4 (hardback)
1. Iron and steel bridges--Design and construction--Examinations--Study guides. 2. Load factor design--Examinations--Study guides. 3. Structural engineering--Examinations--Study guides. I. Kim, Jai B., 1934-

TG416.S56 2013
624.2'52--dc23 2012050940

Visit the Taylor & Francis Web site at
http://www.taylorandfrancis.com

and the CRC Press Web site at
http://www.crcpress.com

Contents

List of Figures

List of Tables

Preface

We have developed this first edition of *LRFD Bridge Design* to comply with the fifth edition of *AASHTO LRFD Bridge Design Specification* (2010) (referred to as AASHTO throughout this book), which should be at your side while using this guide. This book primarily serves engineers studying to take the National Council of Examiners for Engineering and Surveying (NCEES) structural PE examination and the structural depth section of the NCEES civil PE exam. It is also suitable as a classroom text for civil engineering seniors and graduate students, as well as a reference book for practicing engineers.

Acknowledgments

We wish to acknowledge Bucknell University, located in Lewisburg, Pennsylvania, for providing a conducive environment for the writing of this book, and the students in our CENG 461/661 LRFD Bridge Design for their comments and suggestions. Thanks in particular to students Craig Stodart, EIT and Dale Statler, EIT. Also those former students who made significant contributions to this book are Eric J. Weaver, Dave M. Mante, and Jonathan R. Eberle.

Finally we dedicate this book to Yung J. Kim, mother of Robert and wife of Jai B. Kim, who provided the necessary support and motivation for this book.

Should you find an error here, we hope two things happen: first, that you will let us know about it; and second, that you will learn something from the error; we know we will! We would appreciate constructive comments, suggestions for improvement, and recommendations for expansion.

Good luck on the exam!

Introduction

The primary function of *Simplified LRFD Bridge Design* is to serve as a study reference for practicing engineers and students preparing to take the National Council of Examiners for Engineering and Surveying (NCEES) civil and structural exams. As such, this book guides you through the application of the fifth (2010) edition of the *AASHTO LRFD Bridge Design Specifications*, which you must have at your side as you work this book's problems.

Be aware that although AASHTO is incorporated into many major building codes and structural specifications, there may be codes and specifications that differ from, and take priority over, the specifications in AASHTO. In practice you should check with the governing jurisdiction to confirm which codes and specifications must be followed. In addition to AASHTO, you may need to consult other references for more comprehensive explanations of bridge design theory.

This book's first chapter, "LRFD Method of Bridge Design," introduces you to the key steps of LRFD bridge design as they relate to the book's eight design examples and three practice problems. The chapter also includes and describes the use of many key tables and figures from AASHTO. Because this book covers various AASHTO subjects, you may use it to brush up on a few specific subjects, or may study the book in its entirety. Do note, however, that the eight design examples are the most exhaustive in their applications of AASHTO subjects, and that the three practice problems that follow build on concepts and information that have been set out in those first eight examples. You can use this book most effectively by studying the design examples in order. Furthermore, the book's explanations are meant to explain and clarify AASHTO; however, they assume that the reader can refer directly to AASHTO itself when necessary. Among the book's examples are references to AASHTO tables ("A Tbl . . ."), sections ("A Sec . . ."), figures ("A Fig . . ."), and equations ("A Eq . . .").

Throughout the book, example and practice problems illustrate How To Use the *AASHTO LRFD Bridge Design Specifications*, fifth edition (2010). Take your time with these and make sure you understand each example before moving ahead. Keep in mind, though, that in actual design situations there are often several correct solutions to the same problem.

If You Are a Practicing Engineer, Engineering Student, or Instructor

Although this book is primarily intended to aid in exam preparation, it is also a valuable aid to engineers, and can serve as a classroom text for civil engineering seniors and graduate students. For anyone using this book, the design examples serve as a step-by-step, comprehensive guide to bridge design using AASHTO.

If You Are an Examinee

If you are preparing to take the NCEES civil, or structural PE exam, work all of the examples in this book to prepare yourself on the application of the principles presented. By solving the problems in this book you will have a better understanding of the elements of bridge design that could be part of the problems on the exams. By reviewing the solutions, you will learn efficient problem-solving methods that may benefit you in a timed exam.

About the Exams

In April 2011, the new 16-hour structural exam replaced the separate Structural I and II exams. The new exam is a breadth and depth exam offered in two components on successive days. The eight-hour Vertical Forces (Gravity/Other) and Incidental Lateral component is offered only on Friday and focuses on gravity loads and lateral earth pressures. The eight-hour Lateral Forces (Wind/Earthquake) component is offered only on Saturday and focuses on wind and earthquake loads.

Each component of the SE exam has a breadth (morning) and a depth (afternoon) module. Examinees must take the breadth module of each component and one of the two depth modules in each component.

Breadth modules (morning sessions): These modules contain questions covering a comprehensive range of structural engineering topics. All questions are multiple choice.

Depth modules (afternoon sessions): These modules focus more closely on a single area of practice in structural engineering. Examinees must choose either buildings or bridges. Examinees must work the same topic area on both components.

The civil PE exam consists of two sessions, each lasting four hours and consisting of 40 multiple choice questions, but the questions in the morning and afternoon sessions are of about equal difficulty. The morning (breadth) session of the exam may contain general bridge design–related problems. The structural afternoon (depth) session of the exam may include more in-depth bridge design–related problems. The problems in each session typically require an average of six minutes to work

Although the format of the design examples presented in this book differs from those six-minute problems for the civil and structural PE exams, as you work the problems in this book in preparation for either the civil or structural exam, you will find all of the topics covered here also covered on the structural PE exam in some form or another. Using this book will help you gain a broader knowledge base and understanding of the many bridge design subjects covered on exams.

Editors

Jai B. Kim, PE, PhD, is a professor emeritus of civil and environmental engineering at Bucknell University. He was department chairman for 26 years. Also, currently he is a bridge consultant (since 1980) and president of BKLB Structural Consultants, Inc. Recently he was a structural engineer at the Federal Highway Administration (FHWA). He has been active in bridge research for over 40 years, and is currently a member of the Transportation Research Board Committee of Bridges and Structures. He also served on the Structural PE Exam Committee of the National Council of Examiners and Surveying (NCEES) for many years. He holds a BSCE and MSCE from Oregon State University and a PhD from the University of Maryland.

Robert H. Kim, PE, MSCE, is chief design engineer for BKLB Structural Consultants, Inc. He has extensive experience in the design, research, and construction of highway bridges. He has authored and presented several papers related to bridge engineering. Robert's three books, *Bridge Design for the Civil and Structural Professional Exams,* Second Edition; *Timber Design,* Seventh Edition; and *Civil Discipline Specific Review for the FE/EIT Exam* are well read by both students and engineers. In 2013, he is working on a bridge rehabilitation design in Connecticut. He holds a BS from Carnegie Mellon University and a MSCE from The Pennsylvania State University.

Jonathan R. Eberle, BSCE, is engaged in research with focus on the seismic design and analysis of structures at Virginia Polytechnic Institute and State University as a graduate student. He holds a BSCE from Bucknell University.

Contributors

David M. Mante, BSCE, performed a rigorous full-scale laboratory testing program focused on developing and testing of an innovative concrete bridge deck system. Presently, as a PhD student at Auburn University, he is a guest lecturer in undergraduate civil engineering courses and actively performs research related to prestressed concrete bridge girders. He holds a BSCE from Bucknell University.

Eric J. Weaver, PE, M.ASCE, M.ASME, graduated from Bucknell University with a BS in civil engineering and earned an MEng in structural engineering from Lehigh University, where the primary focus of his research was fatigue and life-cycle analysis of steel truss bridges. Following graduate school, Eric worked for several years as a design engineer on NASA's Space Shuttle program and is currently employed as a structural engineer for Westinghouse Electric Company.

Nomenclature

Symbol: Definition (Units)

A: bearing pad area (in^2)
A: area of stringer, beam, or girder (in^2)
a: depth of equivalent rectangular stress block (in)
A_1: factor for dead load used in computing the rating factor
A_2: factor for live load used in computing the rating factor
A_b: area of concrete reinforcing bar (in^2)
A_c: area of composite section (in^2)
ADT: average daily traffic (vehicles/day)
ADTT: average daily truck traffic
$ADTT_{SL}$: single-lane average daily truck traffic
A_g: gross area of cross-section (in^2)
A_{gc}: area of transformed gross composite section (in^2)
A_{ps}: area of prestressing steel (in^2)
A_s: area of nonprestressed reinforcement (in^2)
A_s: peak seismic ground acceleration coefficient modified by short-period site factor
$A_{s,temp}$: area of temperature reinforcement in concrete slab (in^2)
A_v: area of transverse reinforcement with distance s (in^2)
b: width of beam or width of the compression face of the member (in)
b_c: width of the compression flange (in)
b_e: effective flange width for beams (in)
b_{et}: transformed effective deck width (in)
b_f: full width of the flange (in)
b_i: flange width of interior beam (in)
b_{min}: minimum width of T-beam stem (in)
BR: vehicular braking force (kips)
BR: vertical braking force (kips/ft)
BR_{hor}: horizontal braking force at the top of the abutment (kips/ft)
BR_{max}: maximum braking force (kips)
BR_{tandem}: braking force resulting from tandem, single traffic lane (kips)
$BR_{tandem+lane}$: braking force resulting from tandem and lane load, single traffic lane (kips)
BR_{truck}: braking force resulting from truck, single traffic lane (kips)

$BR_{truck+lane}$: braking force resulting from truck and lane load, single traffic lane (kips)
BR_{vert}: vertical braking force at the top of the abutment (kips/ft)
b_s: effective width of concrete deck (in)
b_s: width of beam (in)
$b_{s,ext}$: effective flange width for exterior beams (in)
$b_{s,int}$: effective flange width for interior beams (in)
b_t: width of the tension flange (in)
b_f: flange width of steel beam section (in)
b_v: width of web (in)
BW: barrier weight (kips/ft)
b_w: web width (in)
c: distance from the extreme compression fiber to the neutral axis (in)
C: ratio of the shear buckling resistance to the shear specified minimum yield strength
C: stiffness parameter
C&P: curb and parapet cross-section area (ft^2)
c.g.: center of gravity
CE: vehicular centrifugal force
CL: center line
CR: forces resulting from creep
C_{rb}: distance from top of concrete deck to bottom layer of longitudinal concrete deck reinforcement (in)
C_{rt}: distance from top of concrete deck to top layer of longitudinal concrete deck reinforcement (in)
CT: vehicular collision force
CV: vessel collision force
D: clear distance between flanges (in)
D: dead load (lbf)
D: depth of steel beam (in)
D: width of distribution per lane (ft)
d: depth of beam or stringer (in)
d_b: nominal diameter of reinforcing bar, wire, or prestressing strand (in)
d_c: concrete cover measured from extreme tension fiber to the center of the flexural reinforcement located closest thereto (in)
d_c: distance from the compression flange to the PNA (in)
DC: dead load of structural components and nonstructural attachments (kips)
DC_1: noncomposite dead load (kips/ft)
DC_2: composite dead load (kips/ft)
$DC_{C\&P}$: distributed load resulting from curb and parapet self-weight (kips/ft)
DC_{haunch}: noncomposite dead load resulting from haunch self-weight (kips/ft)
D_{cp}: depth of girder web in compression at the plastic moment (in)
DC_{slab}: noncomposite dead load resulting from slab self-weight (kips/ft)
$DC_{stay\text{-}in\text{-}place\ forms}$: noncomposite dead load resulting from self-weight of stay-in-place forms (kips/ft)

$DC_{T\text{-}beam}$: distributed load resulting from T-beam self-weight (kips/ft)
DD: downdrag load
d_e: effective depth from extreme compression fiber to the centroid of the tensile force in the tensile reinforcement (in)
de: horizontal distance from the centerline of the exterior web of exterior beam at the deck level to the interior edge of curb at barrier.
DF: distribution factor for moment or shear
$DF_{deflection}$: distribution factor for deflection
DFM: distribution factor for moment
DFM^E_{fat}: load distribution for fatigue moments, exterior girder
DFM_{ext}: load distribution for moments, exterior girders
$DFM_{fat,ext}$: load distribution for fatigue moments, exterior girder
$DFM_{fat,int}$: load distribution for fatigue moments, interior girder
$DFM_{fatigue}$: load distribution for fatigue moments
DFM^I_{fat}: load distribution for fatigue moments, interior girder
DFM_{int}: load distribution for moments, interior girders
DFM_{me}: distribution factor for moment for multiple design lanes loaded for exterior beams
DFM_{mi}: distribution factor for moment for multiple design lanes loaded for interior beams
DFM_{se}: distribution factor for moment for a single design lane loaded for exterior beams
DFM_{si}: distribution factor for moment for a single design lane loaded for interior beams
DFV: distribution factor for shear
DFV_{ext}: load distribution for shears, exterior girders
$DFV_{fat,ext}$: load distribution for fatigue shears, exterior girder
$DFV_{fat,int}$: load distribution for fatigue shears, interior girder
DFV_{int}: load distribution for shears, interior girders
DFV_{me}: distribution factor for shear for multiple design lanes loaded for exterior beams
DFV_{mi}: distribution factor for shear for multiple design lanes loaded for interior beams
DFV_{se}: distribution factor for shear for a single design lane loaded for exterior beams
DFV_{si}: distribution factor for shear for a single design lane loaded for interior beams
d_{girder}: depth of girder (in)
d_o: transverse stiffener spacing (in)
d_p: distance from extreme compression fiber to the centroid of the prestressing tendons (in)
D_p: distance from the top of concrete deck to the neutral axis of the composite section (in)
d_s: distance from extreme compression fiber to the centroid of the nonprestressed tensile reinforcement (in)

d_s: thickness of concrete deck slab (in)
D_t: depth of the composite section (in)
d_t: distance from the tension flange to the PNA (in)
d_v: effective shear depth (in)
d_w: distance from the web to the PNA (in)
DW: superimposed dead load (wearing surfaces and utilities) (kips or kips/ft)
DW_{FWS}: future wearing surface dead load (kips/ft)
e: correction factor for load distribution for exterior beams
E: modulus of elasticity of steel (ksi)
E_B: modulus of elasticity of beam material (kips/in^2)
E_{beam}: modulus of elasticity of beam (ksi)
E_c: modulus of elasticity of concrete (ksi)
e_c: strand eccentricity at midspan (in)
E_{cg}: modulus of elasticity of concrete after 28 days (ksi)
E_{ci}: modulus of elasticity of concrete at transfer (ksi)
E_{cs}: modulus of elasticity of concrete after losses (ksi)
E_D: modulus of elasticity of deck material (kips/in^2)
E_{deck}: modulus of elasticity of the deck (ksi)
e_g: distance between the centers of gravity of the beam and deck (in)
EH: horizontal earth pressure load
EL: accumulated locked-in force effects resulting from the construction process, including the secondary forces from posttensioning
e_m: average eccentricity at midspan (in)
E_p: modulus of elasticity of prestressing tendons (ksi)
EQ: forces resulting from earthquake loading (kips)
EQ_h: horizontal earthquake loading at the top of the abutment (kips/ft)
ES: earth surcharge load
E_s: modulus of elasticity of prestressing steel (kips/in^2)
E_s: modulus of elasticity of steel (ksi)
f: bending stress (kips/in^2)
f'_c: compressive strength of concrete at 28 days (ksi)
$f'_{c, beam}$: beam concrete strength (kips/in^2)
$f'_{c, deck}$: deck concrete strength (kips/in^2)
f'_{cg}: compressive strength of concrete at 28 days for prestressed I-beams (ksi)
f'_{cgp}: the concrete stress at the center of gravity of prestressing tendons due to prestressing force immediately after transfer and self-weight of member at section of maximum moment (ksi)
f'_{ci}: compressive strength of concrete at time of prestressing transfer (ksi)
f'_{cs}: compressive strength of concrete at 28 days for roadway slab (ksi)
f'_s: stress in compression reinforcement (ksi)
f_{bt}: amount of stress in a single strand at 75% of ultimate stress (kips/in^2)
f_{bu}: required flange stress without the flange lateral bending
f_c: compressive stress in concrete at service load (ksi)
f_{cgp}: concrete stress at the center of gravity of prestressing tendons that results from the prestressing force at either transfer or jacking and the self-weight of the member at sections of maximum moment (ksi)

$\mathbf{f_{ci}}$: temporary compressive stress before losses due to creep and shrinkage (ksi)

$\mathbf{f_{cpe}}$: compressive stress in concrete due to effective prestress forces only (after allowance for all prestress losses) at extreme fiber of section where tensile stress is caused by externally applied loads (ksi)

$\mathbf{f_{cs}}$: compressive strength of concrete after losses (ksi)

$\mathbf{f_{DC}}$: steel top flange stresses due to permanent dead loads (kips/in^2)

$\mathbf{f_{DW}}$: steel top flange stresses due to superimposed dead load (kips/in^2)

$\mathbf{f_f}$: flange stress due to the Service II loads calculated without consideration of flange lateral bending (ksi)

$\mathbf{f_f}$: allowable fatigue stress range (ksi)

$\mathbf{f_{gb}}$: tensile stress at bottom fiber of section (kips)

$\mathbf{f_l}$: flange lateral bending stress due to the Service II loads (ksi)

$\mathbf{f_{LL+IM}}$: steel top flange stresses due to live load including dynamic load allowance (kips/in^2)

$\mathbf{f_{min}}$: minimum live load stress resulting from the fatigue load combined with the permanent loads; positive if in tension (kips/in^2)

$\mathbf{f_{pbt}}$: stress in prestressing steel immediately prior to transfer (ksi)

$\mathbf{f_{pc}}$: compressive stress in concrete (after allowance for all prestress losses) at centroid of cross-section resisting externally applied loads (ksi)*

$\mathbf{f_{pe}}$: compressive stress in concrete due to effective prestress forces only (after allowance for all prestress losses) at extreme fiber of section where tensile stress is caused by externally applied loads (ksi)

$\mathbf{f_{pga}}$: seismic site factor

$\mathbf{f_{ps}}$: average stress in prestressing steel at the time for which the nominal resistance of member is required (ksi)

$\mathbf{f_{pt}}$: stress in prestressing steel immediately after transfer (ksi)

$\mathbf{f_{pu}}$: specified tensile strength of prestressing steel (ksi)

$\mathbf{f_{pul}}$: stress in the strand at the strength limit state (ksi)

$\mathbf{f_{py}}$: yield strength of prestressing steel (ksi)

$\mathbf{f_r}$: modulus of rupture of concrete (psi)

$\mathbf{f_s}$: stress in the mild tension reinforcement at the nominal flexural resistance (ksi)

$\mathbf{f_s}$: stress in the reinforcement (ksi)

$\mathbf{f_s}$: stress in the reinforcement due to the factored fatigue live load (kips/in^2)

$\mathbf{f_{se}}$: effective steel prestress after losses (ksi)

$\mathbf{f_{si}}$: allowable stress in prestressing steel (ksi)

$\mathbf{f_{ss}}$: tensile stress in mild steel reinforcement at the service limit state (ksi)

$\mathbf{f_t}$: excess tension in the bottom fiber due to applied loads (kips)

$\mathbf{f_t}$: tensile stress at the bottom fiber of the T-beam (kips/in^2)

$\mathbf{f_{ti}}$: temporary tensile stress in prestressed concrete before losses (ksi)

$\mathbf{f_{ts}}$: tensile strength of concrete after losses (psi)

FWS: future wearing surface (in)

* In a composite member, f_{pc} is resultant compressive stress at centroid of composite section.

f_y: specified minimum yield strength of reinforcing bars (ksi)
F_y: specified minimum yield strength of steel (ksi)
F_{yc}: specified minimum yield strength of the compression flange (kips/in^2)
F_{yf}: specified minimum yield strength of a flange (ksi)
F_{yt}: specified minimum yield strength of the tension flange (kips/in^2)
F_{yw}: specified minimum yield strength of a web (ksi)
g: centroid of prestressing strand pattern (in)
g: distribution factor
G: shear modulus of bearing pad elastomers (ksi)
$g_{interior}$ = DFV_{mi}: distribution factor designation for interior girders
g_M^{ME}: distribution factor for moment with multiple lanes loaded, exterior girder
g_M^{MI}: distribution factor for moment with multiple lanes loaded, interior girder
g_M^{SE}: distribution factor for moment with single lane loaded, exterior girder
g_M^{SI}: distribution factor for moment with single lane loaded, interior girder
g_V^{ME}: distribution factor for shear with multiple lanes loaded, exterior girder
g_V^{MI}: distribution factor for shear with multiple lanes loaded, interior girder
g_V^{SE}: distribution factor for shear with single lane loaded, exterior girder
g_V^{SI}: distribution factor for shear with single lane loaded, interior girder
H: average annual ambient relative humidity (%)
h: depth of deck (in)
h: overall depth or thickness of a member (in)
H_{contr}: load due to contraction (kips)
h_{min}: minimum depth of beam including deck thickness (in)
$h_{parapet}$: height of parapet (in)
H_{rise}: load due to expansion (kips)
$H_{temp\ fall}$: horizontal force at the top of the abutment due to temperature fall (kips/ft)
$H_{temp,fall}$: horizontal load due to temperature fall (kips/ft)
H_u: ultimate load due to temperature (kips)
I: moment of inertia (in^4)
I: live load impact factor
I_c: composite section moment of inertia (in^4)
I_g: moment of inertia of gross concrete section about centroidal axis, neglecting reinforcement (in^4)
IM: dynamic load allowance
I_p: polar moment of inertia (in^4)
I_x: moment of inertia with respect to the x-axis (in^4)
I_y: moment of inertia with respect to the y-axis (in^4)
I_{yc}: moment of inertia of the compression flange of the steel section about the vertical axis in the plane of the web (in^4)
I_{yt}: moment of inertia of the tension flange of the steel section about the vertical axis in the plane of the web (in^4)
k: shear-buckling coefficient for webs

K_g: longitudinal stiffness parameter (in^4)
L: span length of beam (ft)
LL: vehicular live load, TL + LN
LN: design lane load
LS: live load surcharge
M: bending moment about the major axis of the cross-section (in-kips)
M: moment designation
m: multiple presence factor
$M_{all, inv}$: allowable bending moment for inventory rating (ft-kips)
$M_{all, opr}$: allowable bending moment for operating rating (ft-kips)
M_C: moment at midspan
M_{cr}: cracking moment (in-kips)
M_D: moment due to slab dead load
M_{DC}: moment due to superstructure dead load (ft-kips)
$M_{DC, tot}$: moment for the total component dead load (kips)
M_{DC1}: unfactored moment resulting from noncomposite dead loads (ft-kips)
M_{DC2}: unfactored moment resulting from composite dead loads (ft-kips)
M_{DW}: moment due to superimposed dead load (ft-kips)
$M^E_{U, fat}$: factored fatigue design live load moment, exterior beam (ft-kips)
M^E_{fat+IM}: unfactored distributed fatigue live load moment with impact, exterior beam (ft-kips)
M_f: moment per lane due to fatigue load (in-kips)
$M_{F,fatigue}$: factored moment per beam due to FatigueI load (in-kips)
$M_{fat,ext}$: unfactored distributed moment resulting from fatigue loading, exterior girder (ft-kips)
$M_{fat,int}$: unfactored distributed moment resulting from fatigue loading, interior girder (ft-kips)
$M_{fat,LL}$: fatigue moment due to live load (ft-kips)
$M_{fatigue}$: unfactored moment per beam due to fatigue load (in-kips)
M_g: midspan moment due to beam weight (in-kips)
$mg_{SI,M}$: distribution of live load moment per lane with one design lane loaded for interior beams
MI: multiple lane, interior designation
mi, MI: two or more design lanes loaded, interior girder
$M^I{}_{fat+IM}$: unfactored distributed fatigue live load moment with impact, interior beam (ft-kips)
$M^I{}_{U, fat}$: factored fatigue design live load moment, interior beam (ft-kips)
M_{LL+IM}: total live load moment per lane including impact factor (ft-kips)
M_{ln}: lane load moment per lane (in-kips)
M_{LN}: unfactored live load moment per beam due to lane load (in-kips)
M_{max}: maximum dead load moment (ft-kips)
M_n: nominal flexural resistance (in-kips)
M_p: plastic moment capacity of steel girder (ft-kips)
M_r: factored flexural resistance of a section in bending, ΦM_n (in-kips)
M_s: moment due to superimposed dead loads (ft-kips)

$M_{service}$: total bending moment resulting from service loads
M_{tandem}: tandem load moment per lane (ft-kips)
M_{TL}: unfactored live load moment per beam due to truck load (in-kips)
M_{tr}: HS-20 truck load moment per lane (in-kips)
M_u: factored design moment at section $\leq \Phi M_n$ (in-kips)
n: modular ratio = E_s/E_c or E_p/E_c
N: number of stress cycles over fatigue design life
n: number of stress cycles per truck passage
N_b: number of beams, stringers, or girders
N_c: number of cells in a concrete box girder
N_g: number of girders
N_L: number of design lanes
p: fraction of truck traffic in a single lane
P: total nominal shear force in the concrete deck for the design of the shear connectors at the strength limit state (kips)
PB: base wind pressure corresponding to a wind speed of 100 mph
P_B: base wind pressure specified in AASHTO (kips/ft^2)
P_c: plastic force in the compression flange (kips)
$P_{C\&P}$: load for the curb and parapet for exterior girders (kips/ft)
P_D: design wind pressure (kips/ft^2)
P_e: effective prestress after losses (kips)
PGA: peak seismic ground acceleration coefficient on rock (Site Class B)
P_i: initial prestress force (kips)
PL: pedestrian live load
PNA: plastic neutral axis
P_{pe}: prestress force per strand after all losses (kips)
P_{pi}: prestress force per strand before transfer (kips)
P_{pt}: prestress force per strand immediately after transfer (kips)
P_{rb}: plastic force in the bottom layer of longitudinal deck reinforcement (kips)
P_{rt}: plastic force in the top layer of longitudinal deck reinforcement (kips)
P_s: plastic force in the slab (kips)
P_t: plastic force in the tension flange (kips)
P_w: plastic force in the web (kips)
Q: total factored load (kips)
Q_i: force effect
Q_i: force effect from various loads
R: reaction at support (kips)
R_F: rating factor for the live load carrying capacity
RF: rating factor for the live load carrying capacity
R_h: hybrid factor
R_n: nominal resistance
R_r: factored resistance (ΦR_n)
RT: load rating for the HS-20 load at the inventory level (tons)
S: section modulus of section (in^3)
s: spacing of bars or stirrups (in)

S: spacing of beams or webs (ft)
S: spacing of supporting elements (ft)
S_b, S_t: noncomposite section moduli (in^3)
S_{bc}, S_{tc}: section moduli of composite beam section at the bottom and top extreme fibers, respectively (in^3)
S_{bottom}: section modulus of the bottom steel flange (in^3)
S_c: section modulus for the extreme fiber of the composite section where tensile stress is caused by externally applied loads (in^3)
S_c, S_{bc}: composite section moduli where the tensile stress is caused by externally applied loads (in^3)
S_e: effective span length (ft)
SE: loads resulting from settlement
SE: single lane, exterior designation
se, SE: single design lane loaded, exterior girder
S_g: section modulus for gross section
SH: loads resulting from shrinkage
SI: single lane, interior designation
si, SI: single design lane loaded, interior girder
s_{max}: maximum spacing of flexural reinforcement (in)
S_{nc}: section modulus for the extreme fiber of the monolithic or noncomposite section where tensile stress is caused by externally applied loads (in^3)
$S_{nc,bottom}$: section modulus for extreme bottom fiber of the monolithic or noncomposite section where tensile stress is caused by externally applied loads (in^3)
$S_{nc,top}$: section modulus for extreme top fiber of the monolithic or noncomposite section where compressive stress is caused by externally applied loads (in^3)
S_{ncb}: section modulus for extreme bottom fiber of the monolithic or noncomposite section where tensile stress is caused by externally applied loads (in^3)
S_{nct}: section modulus for extreme top fiber of the monolithic or noncomposite section where compressive stress is caused by externally applied loads (in^3)
S_{top}: section modulus for the top flange (in^3)
S_x: section modulus with respect to the x-axis (in^3)
S_{xt}: elastic section modulus about the major axis of the section to the tension flange (in^3)
S_{xx}: section modulus with respect to the y-axis (in^3)
S_y: section modulus with respect to the y-axis (in^3)
t: slab thickness (in)
$t_{bearing}$: thickness of bearing (in)
t_c: thickness of a compression flange (in)
t_d: deck thickness (in)
t_{deck}: thickness of deck (in)
t_f: flange thickness (in)

t_g: depth of steel girder or corrugated steel plank including integral concrete overlay or structural concrete component, less a provision for grinding, grooving, or wear (in)
TG: loads resulting from temperature gradient
TL: design truck load, or design tandem load
t_{min}: minimum depth of concrete slab to control deflection (in)
t_o: depth of structural overlay (in)
t_s: thickness of concrete slab (in)
t_t: thickness of the tension flange (in)
TU: loads resulting from uniform temperature
t_w: web thickness (in)
U: factored force effect
V: shear designation
V: shear force (kips)
V_B: base design wind velocity (mph)
V_c: shear resistance provided by the concrete (kips)
V_{cr}: shear-buckling resistance (kips)
V_{DC}: shear due to superstructure dead load (kips)
$V_{DC, tot}$: shear for the total component dead load (kips)
V_{DC1}: unfactored shear resulting from noncomposite dead loads (kips)
V_{DC2}: unfactored shear resulting from composite dead loads (kips)
V_{DL}: unfactored shear force caused by DL (kips)
V_{DW}: shear due to superimposed dead load (kips)
V_{DZ}: design wind velocity (mph)
V_{fat}: shear force resulting from fatigue load (kips)
$V_{fatigue}$: fatigue load shear per lane (kips)
$V_{fatigue+IM}$: fatigue load shear per girder (kips)
V_{LL+IM}: total live load shear per lane including impact factor (ft-kips)
V_{ln}: lane load shear per lane (kips)
V_{LN}: unfactored live load shear per beam due to lane load (kips)
V_{max}: maximum dead load shear (kips)
V_n: nominal shear resistance (kips)
V_p: plastic shear resistance of the web (kips)
V_p: shear yielding of the web (kips)
$V_{permanent}$: shear due to unfactored permanent load (kips)
V_s: shear resistance provided by shear reinforcement (kips)
V_{tandem}: tandem load shear per lane (kips)
V_{TL}: unfactored live load shear per beam due to truck load (kips)
V_{tr}: truck load shear per lane (kips)
V_u: factored shear force at section (kips)
v_u: average factored shear stress on concrete (ksi)
$V_{u,ext}$: factored shear force at section in external girder (kips)
$V_{u,int}$: factored shear force at section in internal girder (kips)
$V_{u,total}$: total factored shear force at section (kips)
w: distributed load (kips/ft^2)

W: weight in tons of truck used in computing live load effect
w: width of clear roadway (ft)
WA: water load and stream pressure
$\mathbf{w_c}$: self-weight of concrete (kips/ft^3)
$\mathbf{w_{C\&P}}$: distributed load resulting from self-weight of curb and parapet (kips/ft)
$\mathbf{w_{DC}}$: distributed load of weight of supported structure (kips/ft^2)
$\mathbf{w_{DW}}$: distributed load of superimposed dead load (kips/ft^2)
$\mathbf{w_{FWS}}$: future wearing surface load (kips/ft^2)
WL: loads resulting from wind forces on live load
WL: wind pressure on vehicles, live load
$\mathbf{WL_h}$: horizontal loading due to wind pressure on vehicles
$\mathbf{WL_h}$: horizontal wind loading at the top of the abutment (kips/ft)
$\mathbf{WL_v}$: vertical wind loading at the top of the abutment (kips/ft)
$\mathbf{w_s}$: superimposed dead loads, parapet/curb load plus the future wearing surface load (kips/ft^2)
WS: wind load pressure on superstructure
WS: wind pressures on superstructures (kips)
$\mathbf{WS_h}$: horizontal load on top of abutment due to wind pressure on superstructure
$\mathbf{WS_h}$: horizontal wind loading at the top of the abutment (kips/ft)
$\mathbf{w_{slab}}$: distributed load of concrete slab (kips/ft^2)
$\mathbf{w_{slab,ext}}$: deck slab distributed load acting on exterior girder (kips/ft)
$\mathbf{w_{slab,int}}$: deck slab distributed load acting on interior girder (kips/ft)
$\mathbf{WS_{sub}}$: horizontal wind load applied directly to the substructure
$\mathbf{WS_{total}}$: total longitudinal wind loading (kips)
$\mathbf{WS_v}$: vertical load on top of abutment due to wind pressure on superstructure
$\mathbf{WS_v}$: vertical wind loading along the abutment (kips/ft)
X: distance from load to point of support (ft)
x: distance from beam to critical placement of wheel load (ft)
x: distance of interest along beam span (ft)
$\mathbf{y'_t}$, $\mathbf{y'_b}$, $\mathbf{y_t}$, and $\mathbf{y_b}$: for composite beam cross-section (in)
$\mathbf{y_b}$: distance from the bottom fiber to the centroid of the section (in)
$\mathbf{y_{bs}}$: distance from the center of gravity of the bottom strands to the bottom fiber (in)
$\mathbf{y_t}$: distance from the neutral axis to the extreme tension fiber (in)
$\mathbf{y_t}$, $\mathbf{y_b}$: distance from centroidal axis of beam gross section (neglecting reinforcement) to top and bottom fibers, respectively (in)
$\mathbf{Z_{req'd}}$: required plastic section modulus (in^3)
$\boldsymbol{\alpha}$: angle of inclination of stirrups to longitudinal axis
$\boldsymbol{\alpha}$: angle of inclination of transverse reinforcement to longitudinal axis (deg)
$\boldsymbol{\beta}$: factor indicating ability of diagonally cracked concrete to transmit tension
$\boldsymbol{\beta_1}$: factor for concrete strength
$\boldsymbol{\beta_1}$: ratio of the depth of the equivalent uniformly stressed compression zone assumed in the strength limit state to the depth of the actual compression zone

$\boldsymbol{\beta}_s$: ratio of the flexural strain at the extreme tension face to the strain at the centroid of the reinforcement layer nearest the tension face
$\boldsymbol{\gamma}$: load factor
γ_e: exposure factor
γ_h: correction factor for relative humidity of the ambient air
γ_i: load factor; a statistically based multiplier applied to force effects including distribution factors and load combination factors
γ_p: load factors for permanent loads
γ_{st}: correction factor for specified concrete strength at the time of the prestress transfer to the concrete
δ: beam deflection (in)
$\boldsymbol{\Delta}_{25\%\ truck}$: 25% of deflection resulting from truck loading (in)
$\boldsymbol{\Delta}_{25\%\ truck+\ lane}$: 25% of deflection resulting from truck loading plus deflection resulting from lane loading (in)
$\boldsymbol{\Delta}_{contr}$: contraction resulting from thermal movement (in)
$\boldsymbol{\Delta}_{contr}$: contractor thermal movement
$\boldsymbol{\Delta}_{exp}$: expansion resulting from thermal movement (in)
$\boldsymbol{\Delta}_{exp}$: expansion thermal movement
$\boldsymbol{\Delta f}_{ext}$: maximum stress due to fatigue loads for exterior girders (kips/in^2)
$\boldsymbol{\Delta f}_{int}$: maximum stress due to fatigue loads for interior girders (kips/in^2)
($\boldsymbol{\Delta f}$): load-induced stress range due to fatigue load (ksi)
($\boldsymbol{\Delta F}$)$_n$: nominal fatigue resistance (ksi)
$\boldsymbol{\Delta f}_{pES}$: sum of all losses or gains due to elastic shortening or extension at the time of application of prestress and/or external loads (ksi)
$\boldsymbol{\Delta f}_{pLT}$: losses due to long-term shrinkage and creep of concrete, and relaxation of the steel (ksi)
$\boldsymbol{\Delta f}_{pR}$: estimate of relaxation loss taken as 2.4 kips/in^2 for low relaxation strand, 10.0 kips/in^2 for stress-relieved strand, and in accordance with manufacturer's recommendation for other types of strand (kips/in^2)
$\boldsymbol{\Delta f}_{pT}$: total loss (ksi)
($\boldsymbol{\Delta F}$)$_{TH}$: constant amplitude (ksi)
$\boldsymbol{\Delta}_{truck}$: deflection resulting from truck loading (in)
δ_{LL}: deflection due to live load per lane (in)
δ_{LL+IM}: deflection due to live load per girder including impact factor (in)
δ_{ln}: deflection due to lane load (in)
δ_{max}: maximum deflection for vehicular load (in)
ε_x: tensile strain in the transverse reinforcement
$\boldsymbol{\eta}$: load modifier
$\boldsymbol{\eta}_D$: ductility factor (strength only)
$\boldsymbol{\eta}_i$: load modifier relating to ductility redundancy, and operational importance = 1.0 (for conventional designs)
$\boldsymbol{\eta}_I$: operational importance factor (strength and extreme only) = 1.0 for (for conventional bridges)
$\boldsymbol{\eta}_R$: redundancy factor
$\boldsymbol{\theta}$: angle of inclination of diagonal compressive stresses (degrees)

$\mathbf{\Phi}$: resistance factor
$\mathbf{\Phi}_c$: condition factor
$\mathbf{\Phi}_f$: resistance factor for flexure
$\mathbf{\Phi}_s$: system factor
$\mathbf{\Phi}_v$: resistance factor for shear

1

LRFD Method of Bridge Design

In load and resistance factor design (LRFD), bridges are designed for specific limit states that consider various loads and resistance. These limit states include strength, extreme event, service, and fatigue, and are defined in the first section of this chapter. Subsequent sections cover the following load and resistance factors in more detail:

- Load combinations and load factors
- Strength limit states for superstructure design
- Resistance factors for strength limits
- Design live loads
- Number of design lanes
- Multiple presence of live loads
- Dynamic load allowances
- Live load distribution factors
- Load combinations for the Strength I Limit State
- Simple beam moments and shears carrying moving concentrated loads

Limit States

The following load combinations are defined in AASHTO. Bridges are designed for these limit states with consideration for the load and resistance factors detailed in later sections of this chapter.

A Art. 3.4.1, 1.3.2*

- **Strength I**: Basic load combination related to the normal vehicular use of the bridge without wind
- **Strength II**: Load combination relating to the use of the bridge by owner-specified special design vehicles and/or evaluation permit vehicles, without wind

* The article numbers in the *2010 Interim Revisions* to the *AASHTO Bridge Design Specifications*, fifth edition, 2010 are by the letter A if specifications and letter C or Comm if commentary.

- **Strength III**: Load combination relating to the bridge exposed to wind velocity exceeding 55 mph without live loads
- **Strength IV**: Load combination relating to very high dead load to live load force effect ratios exceeding about 7.0 (e.g., for spans greater than 250 ft)
- **Strength V**: Load combination relating to normal vehicular use of the bridge with wind velocity of 55 mph
- **Extreme Event I**: Load combinations including earthquake
- **Extreme Event II**: Load combinations relating to ice load or collisions by vessels and vehicles
- **Service I**: Load combination relating to the normal operational use of the bridge with 55 mph wind. Also used for live load deflection control

Art. 2.5.2.6.2

- **Service II**: Load combination intended to control yielding of the steel structures and slip of slip-critical connections due to vehicular live load
- **Service III**: Load combination relating only to tension in prestressed concrete structures with the objective of crack control
- **Fatigue I**: Fatigue and fracture load combination related to infinite load-induced fatigue life
- **Fatigue II**: Fatigue and fracture load combination related to finite load-induced fatigue life

The following terms are defined for limit states:

γ_i = Load factor: a statistically based multiplier applied to force effects

ϕ = Resistance factor: a statistically based multiplier applied to nominal resistance, as specified in AASHTO Specification Sections 5–8, and 10–12

η_i = Load modifier: a factor relating to ductility, redundancy, and operational importance

η_D = A factor relating to ductility, as specified in AASHTO Article 1.3.3

η_R = A factor relating to redundancy as specified in AASHTO Article 1.3.4

η_I = A factor relating to operational importance as specified in AASHTO Article 1.3.5

Q_i = Force effect

R_n = Nominal resistance

R_r = Factored resistance: (ϕR_n)

Effects of loads must be less than or equal to the resistance of a member (or its components), or $\eta\gamma Q \le \Phi R_n = R_r$

A Art. 1.3.2

$$\Sigma \eta_i \gamma_i Q_i \le \phi R_n.$$

A Eq. 1.3.2.1-1

For loads for which a maximum value of γ_i is appropriate,

$$\eta_i = \eta_D \eta_R \eta_I \ge 0.95$$

A Eq. 1.3.2.1-2

For loads for which a minimum value of γ_i is appropriate,

$$\eta_i = \frac{1}{\eta_D \eta_R \eta_I} \le 1.0$$

A Eq. 1.3.2.1-3

Load Combinations and Load Factors

Please see Tables 1.1 (**A Tbl. 3.4.1-1**) and 1.2 (**A Tbl. 3.4.1-2**), which show the load factors for various load combinations and permanent loads.

Loads and Load Designation

A Art. 3.3.2

The following permanent and transient loads and forces shall be considered in bridge design:

Permanent Loads

CR = force effects due to creep
DD = downdrag force
DC = dead load of structural components and nonstructural attachments
DW = dead load of wearing surfaces and utilities
EH = horizontal earth pressure load

TABLE 1.1 (AASHTO Table 3.4.1-1)

Load Combinations and Load Factors

Load Combination Limit State	*DC* *DD* *DW* *EH* *EV* *ES* *EL* *PS* *CR* *SH*	*LL* *IM* *CE* *BR* *PL* *LS*	*WA*	*WS*	*WL*	*FR*	*TU*	*TG*	*SE*	Use One of These at a Time *EQ*	*IC*	*CT*	*CV*
Strength I (unless noted)	γp	1.75	1.00	—	—	1.00	0.50/1.20	γ_{TG}	γ_{SE}	—	—	—	—
Strength II	γp	1.35	1.00	—	—	1.00	0.50/1.20	γ_{TG}	γ_{SE}	—	—	—	—
Strength III	γp	—	1.00	1.40	—	1.00	0.50/1.20	γ_{TG}	γ_{SE}	—	—	—	—
Strength IV	γp	—	1.00	—	—	1.00	0.50/1.20	—	—	—	—	—	—
Strength V	γp	1.35	1.00	0.40	1.0	1.00	0.50/1.20	γ_{TG}	γ_{SE}	—	—	—	—
Extreme Event I	γp	γEQ	1.00	—	—	1.00	—	—	—	1.00	—	—	—
Extreme Event II	γp	0.50	1.00	—	—	1.00	—	—	—	—	1.00	1.00	1.00
Service I	1.00	1.00	1.00	0 30	1.0	1.00	1.00/1.20	γ_{TG}	γ_{SE}	—	—	—	—
Service II	1.00	1.30	1.00	—	—	1.00	1.00/1.20	—	—	—	—	—	—
Service III	1.00	0.80	1.00	—	—	1.00	1.00/1.20	γ_{TG}	γ_{SE}	—	—	—	—
Service IV	1.00	—	1.00	0.70	—	1.00	1.00/1.20	—	1.0	—	—	—	—
Fatigue I— *LL*, *IM*, & *CE* only	—	1.50	—	—	—	—	—	—	—	—	—	—	—
Fatigue II— *LL*, *IM*, & *CE* only	—	0.75	—	—	—	—	—	—	—	—	—	—	—

TABLE 1.2 (AASHTO Table 3.4.1-2)

Load Factors for Permanent Loads, γ_p

Type of Load. Foundation Type, and Method Used to Calculate Downdrag		Load Factor	
		Maximum	Minimum
DC: Component and Attachments		1.25	0.90
DC: Strength IV only		1.50	0.90
DD: Downdrag	Piles, α Tomlinson Method	1.4	0.25
	Piles. λ Method	1.05	0.30
	Drilled shafts. O'Neill and Reese (1999) Method	1.25	0.35
DW: Wearing Surfaces and Utilities		1.50	0.65
EH: Horizontal Earth Pressure			
• Active		1.50	0.90
• At-Rest		1.35	0.90
• *AEP* for Anchored Walls		1.35	N/A
EL: Locked-in Construction Stresses		1.00	1.00
EV: Vertical Earth Pressure			
• Overall Stability		1.00	N/A
• Retaining Walls and Abutments		1.35	1.00
• Rigid Buried Structure		1.30	0.90
• Rigid Frames		1.35	0.90
ES: Earth Surcharge		1.50	0.75

EL = miscellaneous locked-in force effects resulting from the construction process, including jacking apart of cantilevers in segmental construction

ES = earth surcharge load

EV = vertical pressure from dead load of earth fill

PS = secondary forces from posttensioning

SH = force effects due to shrinkage

γp = load factor for permanent loading

Transient Loads

BR = vehicular braking force

CE = vehicular centrifugal force

CT = vehicular collision force

CV = vessel collision force

EQ = earthquake load

FR = friction load

IC = ice load
IM = vehicular dynamic load allowance
LL = vehicular live load
LN = design lane load
LS = live load surcharge
PL = pedestrian live load
SE = force effect due to settlement
TG = force effect due to temperature gradient
TL = design truck load or design tandem load
TU = force effect due to uniform temperature
WA = water load and stream pressure
WL = wind on live load
WS = wind load on structure

Strength Limit States for Superstructure Design

The load effect, Q, is given in the following equations in relation to various limit states:

Strength I: Q = 1.25 DC + 1.5 DW + 1.75 LL; LL = TL + LN
Strength II: Q = 1.25 DC + 1.5 DW + 1.35 LL
Strength III: Q = 1.25 DC + 1.5 DW + 1.4 WS
Strength IV: Q = 1.5 DC + 1.5 DW
Service I: Q = 1.0 DC + 1.0 DW + 1.0 LL + 1.0 WA + 0.3 WS + 1.0 WL
Service II: Q = 1.0 DC + 1.0 DW + 1.3 LL
Service III: Q = 1.0 DC + 1.0 DW + 0.8 LL +1.0 WA
Fatigue I: Q = 1.5 (LL + IM)
Fatigue II: Q = 0.75 (LL + IM)

Resistance Factors, Φ, for Strength Limits

Resistance factors, Φ, for strength limit states are given for various structural categories in AASHTO. Selected resistance factors that are most frequently encountered follow.

For service and extreme event limit states, resistance factors ϕ shall be taken as 1.0, except for bolts.

A Art. 1.3.2.1

Flexure and tension in reinforced concrete	:	0.90

A Art. 5.5.4.2

Flexure and tension in prestressed concrete	:	1.00
Shear in concrete	:	0.90
Axial compression in concrete	:	0.75
Flexure in structural steel	:	1.00

A Art. 6.5.4.2

Shear in structural steel	:	1.00
Axial compression in structural steel	:	0.90
Tension, yielding in gross section	:	0.95

Design Live Load HL-93

HL-93 is a notional live load where H represents the HS truck, L the lane load, and 93 the year in which the design live load HL-93 was adopted.

The design live load designated as the HL-93 consists of a combination of:

A Art. 3.6.1.2

- Design truck (HS-20) or design tandem (a pair of 25 kip axles 4 ft apart),
- Design lane load of 0.64 kip-ft uniformly distributed in the longitudinal direction. Transversely the design lane load is distributed over a 10 ft width within a 12 ft design lane. Note that the design lane load is not subject to dynamic allowance.

For both design truck and design tandem loads, the transverse spacing of wheels is taken as 6.0 ft. The uniform lane load may be continuous or discontinuous as necessary to produce the maximum force effect.

Fatigue Live Load

The fatigue load will be one truck or axles as specified in Figure 1.1 (A Art. 3.6.1.2.2) with a constant spacing of 30 ft between the 32.0 kip axles. The frequency of the fatigue load shall be taken as the single-lane average daily truck traffic. When the bridge is analyzed by approximate load distribution (A Art. 4.6.3), the distribution factor for one traffic lane shall be used. See Figure 1.2.

A Art. 3.6.1.4

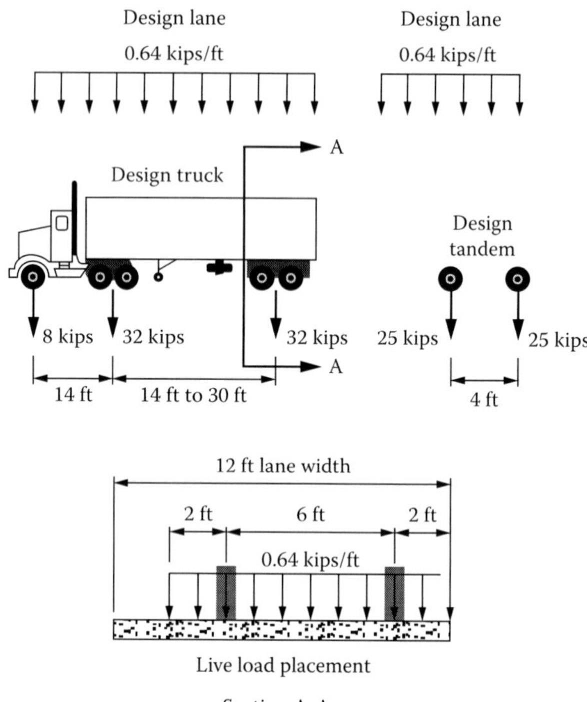

FIGURE 1.1
Design truck (HS-20), design tandem load (a pair of 25 kip axles 4 ft apart), and design lane load (0.64 kips/ft longitudinally distributed). (*Source:* **Art. 3.6.1.2.2.**)

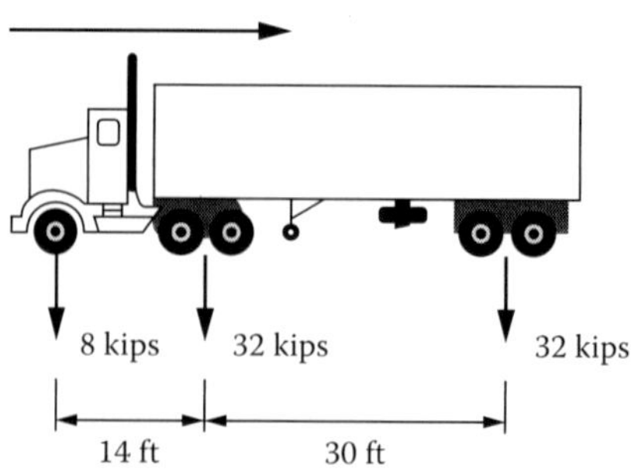

FIGURE 1.2
Fatigue live loading. (*Source:* **Art. 3.6.1.4.**)

Number of Design Lanes, N_L

The number of design lanes, N_L, is the integer portion of the ratio of the clear roadway width (ft), w, and the width of the design traffic lane (12 ft). For example, if the clear roadway width is 40 ft,

A Art. 3.6.1.1.1

$$N_L = \frac{w}{12.0\frac{\text{ft}}{\text{lane}}} = \frac{40\text{ ft}}{12.0\frac{\text{ft}}{\text{lane}}} = 3.33\text{ lanes (3 lanes)}$$

Multiple Presence Factor of Live Load, m

Design trucks will be present in adjacent lanes on roadways with multiple design lanes. Because it is unlikely that three or more adjacent lanes will be loaded simultaneously with trucks, adjustments in design loads are necessary. These factors have been implicitly included in the approximate equations for distribution factors and should be removed for fatigue investigations. See Table 1.3.

A Art 3.6.1.1.2

Therefore for fatigue investigations in which the traffic truck is placed in a single lane, the factor of 1.2 which has been included in the approximate equations should be removed.

A Com. 3.6.1.1.2

The multiple presence m is defined for sites with an ADTT of 5,000 trucks or greater in one direction. For sites with a lower ADTT, reduce force effects by:

TABLE 1.3 (AASHTO Table 3.6.1.1.2-1)

Multiple Presence Factors, m

A Art. 3.6.1.1.2

Number of Loaded Lanes	Multiple Presence Factors, m
1	1.20
2	1.00
3	0.85
>3	0.65

$$100 < \text{ADTT} < 1{,}000 \rightarrow 95\%$$

$$\text{ADTT} < 100 \rightarrow 90\%$$

This adjustment is based on the reduced probability of attaining the design event during a 75-year design life with reduced truck volume.

Dynamic Load Allowance, IM

The dynamic load allowance, IM, is applied only to the design truck load, or tandem, not to the design lane load. The static effects of the design truck or tandem shall be increased for dynamic load allowance. Table 1.4 indicates the dynamic load allowance for different components under different limit states.

A Art. 3.6.2

Live Load Distribution Factors

Live load distribution factors in AASHTO are lane-load distributions, not wheel-load distributions as they were in the *AASHTO Standard Specifications for Highway Bridges*, 17th edition, 2002.

A Art. 4.6.2.2: Appendices A–D

The distribution factors are included in several AASHTO articles, and important provisions are in AASHTO Sec. 4.

The live load moment and shear for beams or girders are determined by applying the lane fraction (distribution factor) in AASHTO Art. 4.6.2.2.2 to the moment and shear due to the loads assumed to occupy 10 ft. transversely within a design lane.

AASHTO 3.6.1.2.1

TABLE 1.4 (AASHTO Table 3.6.2.1-1)

Dynamic Load Allowance, IM

Component	IM (%)
Deck joints, all limit states	75
All other components	
Fatigue and fracture limit state	15
All other limit states	33

UNIVERSITIES AT MEDWAY LIBRARY

Distribution factors are most sensitive to beam (girder) spacing. Span length and longitudinal stiffness have smaller influences.

Load-carrying capacity of exterior beams (girders) shall not be less than that of interior beams (girders).

A Art. 2.5.2.7.1

Because of the many algebraically complex expressions and equations associated with live load distribution, they are discussed in detail in example problems.

Load Combinations for the Strength I Limit State

The total factored force effects, Q, shall be taken as

A Arts 3.3.2, 3.4.1, Tbls. 3.4.1-1, 3.4.1-2

$$Q = \Sigma \eta_i \gamma_i Q_i$$

Where:

η_i = load modifier = 1.0
γ_i = load factor

A Eq. 3.4.1-1

For load combination limit state Strength I,

$Q = 1.25\ DC + 1.50\ DW + 1.75\ (TL + LN)$

$M_u = 1.25\ MDC + 1.50\ MDW + 1.75\ (MTL + MLN)$

$V_u = 1.25\ V_{DC} + 1.50\ V_{DW} + 1.75\ (V_{TL} + V_{LN})$

TL = Truck load or tandem load

LN = Lane load

η_D = 1.0 for conventional designs

A Art. 1.3

η_R = 1.0 for conventional levels of redundancy

η_I = 1.0 for typical bridges

Φ = 1.0 for service and fatigue limit states

A 13.2.1; A 5.5.4.2; A 6.5.4.2; A C6.6.1.2.2

ϕ For concrete structures and steel structures, refer to Art. 5.5.4.2.1 and Art. 6.5.4.2, respectively.

Unfactored Dead Load Moments and Shears

M_{DC} = Maximum unfactored moment due to DC
M_{DW} = Maximum unfactored moment due to DW
V_{DC} = Maximum unfactored shear due to DC
V_{DW} = Maximum unfactored shear due to DW

Unfactored Live Load Moments per beam with Distribution Factors DFM, and Dynamic Allowance IM

M_{TL} = (maximum truck or tandem load moment per lane due to design truck load) (DFM) (1 + IM)
M_{LN} = (maximum lane load moment per lane) (DFM)*

Unfactored Live Load Shear per Beam with Distribution Factors DFV, and Dynamic Allowance, IM

V_{TL} = (maximum truck load per shear lane) (DFV) (1+IM)
V_{LN} = (maximum lane load shear per lane) (DFV)[2]

Live Load Distribution Factors for Moment for Beams

DFM_{si} = Single (one) lane loaded for moment in interior beams
DFM_{mi} = Multiple (two or more) lanes loaded for moment in interior beams
DFM_{se} = Single (one) lane loaded for moment in exterior beams
DFM_{me} = Multiple (two or more) lanes loaded for moment in exterior beams

Live Load Distribution Factors for Shear for Beams

DFV_{si} = Single (one) lane loaded for shear in interior beams
DFV_{mi} = Multiple (two or more) lanes loaded for shear in interior beams
DFV_{se} = Single (one) lane loaded for shear in exterior beams
DFV_{me} = Multiple (two or more) lanes loaded for shear in exterior beams

* No impact allowance applies.

Simple Beam Live Load Moments and Shears Carrying Moving Concentrated Loads per Lane

The maximum moment occurs under one of the loads when that load is as far from one support as the center of gravity of all the moving loads on the beam is from the other support. This condition occurs when the center of the span is midway between the center of gravity of the moving loads and the nearest concentrated load where the maximum moment occurs.

The maximum shear due to moving loads occurs at one support when one of the moving loads is at that support. With several moving loads, the location that will produce maximum shear must be determined by trial and error, as shown. Because the maximum moments for uniform loads such as the lane loads and dead loads occur at midspan, the maximum design truck or tandem moment is generally used with the HL-93 center axle at midspan. See Figures 1.3 and 1.4.

Live Load Moments and Shears for Beams (Girders)

A Art. Tbls. 4.6.2.2.2b-1 and 4.6.2.2.2d-1; and 4.6.2.2.3a-1 and 4.6.2.2.3b-1

Live load moments and shears for beams (girders) are determined by applying the distribution factors for live loads per lane in AASHTO tables in 4.4.2.2 to the moment and shears due to the live loads assumed to occupy 10 ft. within a design lane (AASHTO 3.6.1.2.1).

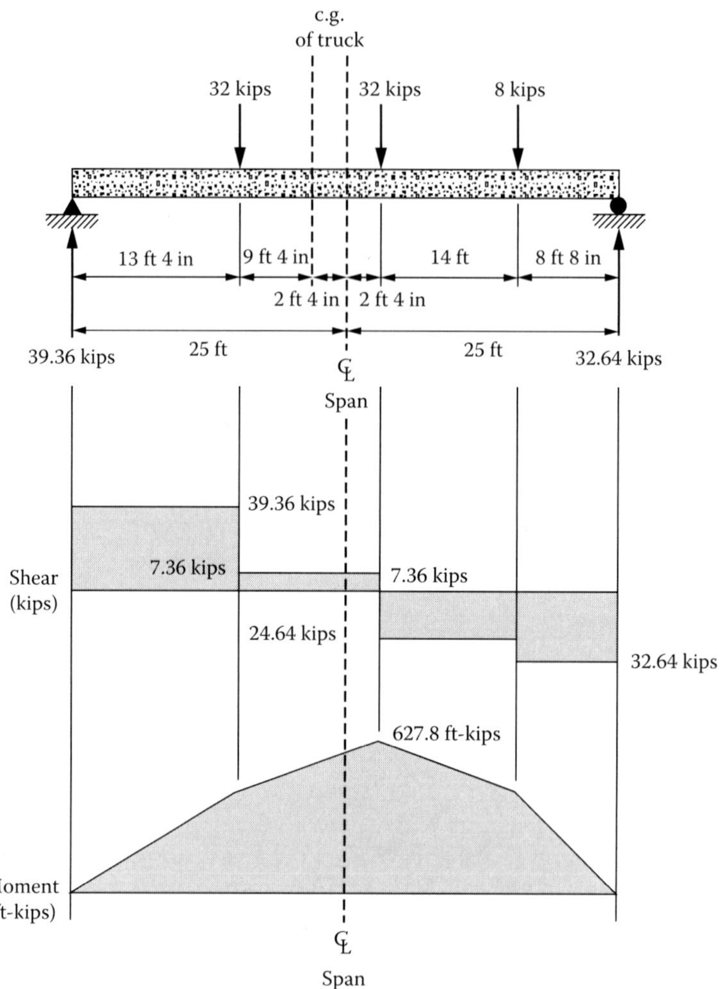

FIGURE 1.3
Shear and moment diagrams for controlling design truck (HS-20) live load position.

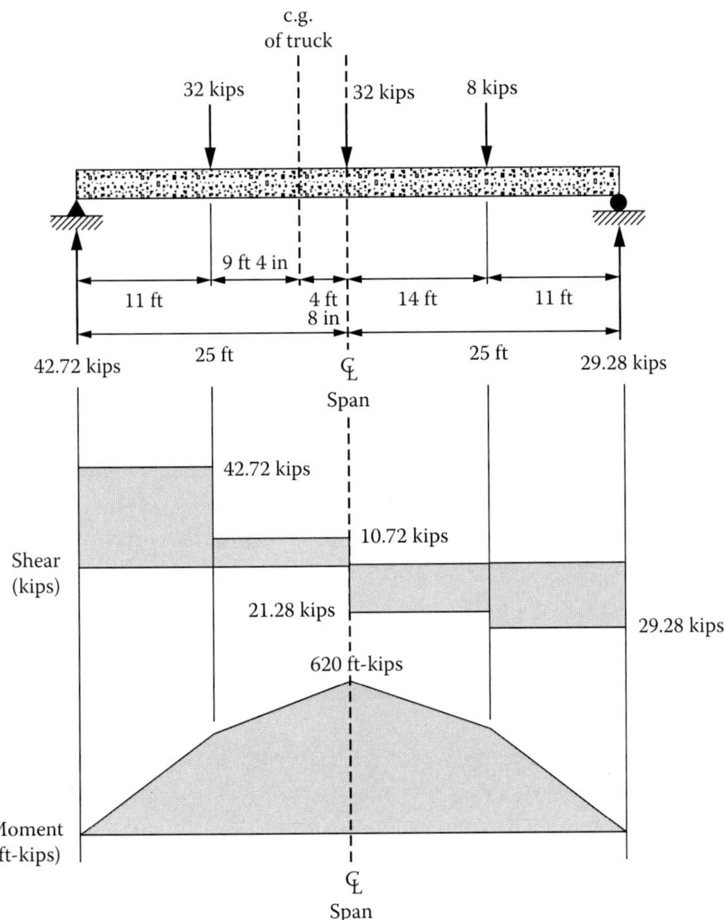

FIGURE 1.4
Shear and moment diagrams for the design truck (HS-20) center axle at midspan.

2

Design Examples

Design Example 1: Reinforced Concrete T-Beam Bridge

Problem Statement

A bridge will be designed with a span length of 50 ft. The superstructure consists of five beams spaced at 10 ft with a concrete deck slab of 9 in. The overall width of the bridge is 48 ft and the clear (roadway) width is 44 ft, 6 in. Design the superstructure of a reinforced, cast-in-place concrete T-beam bridge using the following design specifications.

The three Load Combination Limit States considered are Strength I, Fatigue II, and Service I.

C&P	Curb and parapet cross section	3.37 ft^2
E_c	Modulus of elasticity of concrete	4×10^3 $kips/in^2$
f'_c	Specified compressive strength of concrete	4.5 $kips/in^2$
f_y	Specified yield strength of epoxy-coated reinforcing bars	60 $kips/in^2$
w_c	Self-weight of concrete	0.15 $kips/ft^3$
w_{FWS}	Future wearing surface load	0.03 $kips/ft^2$

Figure 2.1 shows the elevation view, section view, and overhang detail of the reinforced concrete T-beam bridge described

Solution

Step 1: Design T-Beam Using Strength I Limit State

The factored load, Q, is calculated using the load factors given in AASHTO Tables 3.4.1-1 and 3.4.1-2.

$$Q = 1.25\ DC + 1.5\ DW + 1.75(TL+LN)$$

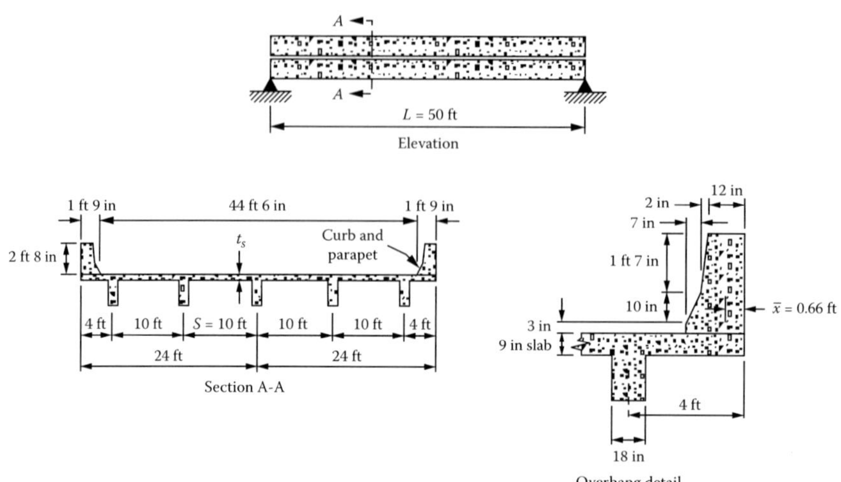

FIGURE 2.1
T-beam design example.

Find the flange width and web thickness.

A Art. 5.14.1.5.1

For the top flange of T-beams serving as deck slabs, the minimum concrete deck must be greater than or equal to 7 in. First, try using a slab with a thickness, t_s, of 9 in.

A Arts. 5.14.1.5.1a, 9.7.1.1

The minimum web thickness is 8 in.

A Art. 5.14.1.5.1c; Com. 5.14.1.5.1c

Minimum concrete cover for main epoxy-coated bars shall be 1 in. Use 1.5 in cover for main and stirrup bars.

A Tbl. 5.12.3-1; Art. 5.12.4

Find the width of T-beam stem (web thickness), b.

A Art. 5.10.3.1.1

The minimum width, b_{min}, is found as follows:

Assume two layers of no. 11 bars for positive reinforcement.

d_b for a no. 11 bar = 1.41 in

Four no. 11 bars in a row and no. 4 stirrup bars require a width of

$$b_{min} = 2\,(2.0 \text{ in cover and no. 4 stirrup}) + 4\,d_b + 3\,(1.5\,d_b)$$

$$= 2\,(2.0 \text{ in}) + 4\,(1.41 \text{ in}) + 3\,(1.5 \times 1.41 \text{ in}) = 16 \text{ in}$$

Try b_w = width of stem = 18 in.

Find the beam depth including deck.

A Tbl. 2.5.2.6.3-1

$$h_{min} = 0.070\ L = 0.070\ (50\ ft \times 12\ in) = 42\ in.$$

Try h = beam width = 44 in.

Find the effective flange width.

A Art. 4.6.2.6.1

Find the effective flange width, where

b_e	effective flange for exterior beams	in
b_i	effective flange width for interior beams	in
b_w	web width	18 in
L	effective span length (actual span length)	50 ft
S	average spacing of adjacent beams	10 ft
t_s	slab thickness	9 in

The effective flange width for interior beams is equal to one-half the distance to the adjacent girder on each side of the girder,

$$b_i = S = 10\ ft \times 12\ in = 120\ in$$

The effective flange width for exterior beams is equal to half of one-half the distance to the adjacent girder plus the full overhang width,

$$b_e = \tfrac{1}{2}\ (10\ ft \times 12\ in) + (4\ ft \times 12\ in) = 108\ in$$

Find the interior T-beam section. Please see Figure 2.2.

The section properties of the preceding interior T-beam are as follows.

The area of the T-beam is

$$A = (9\ in)\ (120\ in) + (35\ in)\ (18\ in) = 1710\ in^2$$

The center of gravity from the extreme tension fiber is

$$y_t = \frac{\Sigma y_t A}{\Sigma A} = \frac{(9\ in)(120\ in)(35\ in + 4.5\ in) + (35\ in)(18\ in)(17.5\ in)}{1710.0\ in^2}$$

$$= 31.4\ in$$

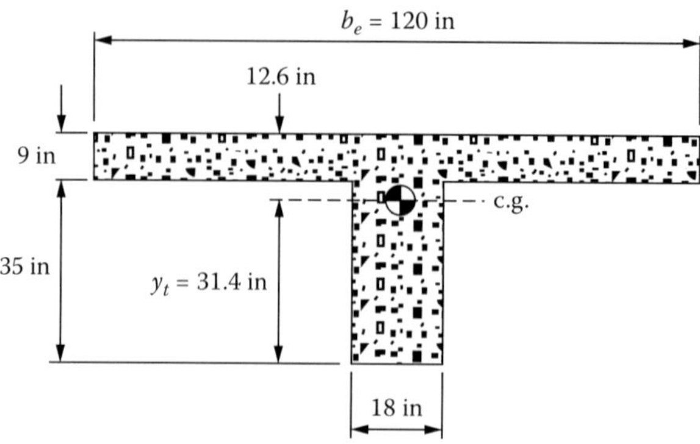

FIGURE 2.2
Interior T-beam section.

The moment of inertia (the gross concrete section) about the center of gravity is

$$I_g = I = \Sigma\left(\bar{I} + Ad^2\right)$$

$$= \frac{(120\text{ in})(9\text{ in})^3}{12} + (120\text{ in})(9\text{ in})(12.5\text{ in} - 4.5\text{ in})^2$$

$$+ \frac{(18\text{ in})(35\text{ in})^3}{12} + (18\text{ in})(35\text{ in})(31.4\text{ in} - 17.5\text{ in})^2$$

$$I_g = 264{,}183.6\text{ in}^4$$

The T-beam stem is

$$(35\text{ in})(18\text{ in}) = 630\text{ in}^2$$

The section modulus at bottom fiber is

$$S = \frac{I_g}{y_t} = \frac{264{,}183.6\text{ in}^4}{31.4\text{ in}} = 8413.5\text{ in}^3$$

Find the number of design lanes.

A Art. 3.6.1.1.1

The number of design lanes, N_L, is the integer portion of the ratio of the clear road width divided by a 12 ft traffic lane width.

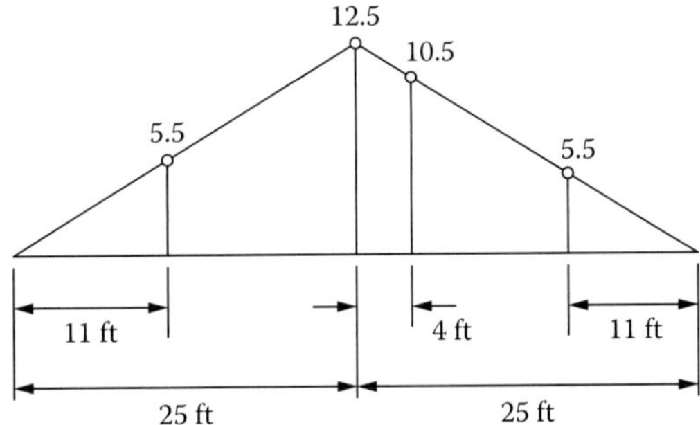

FIGURE 2.3a
Influence lines for moment at midspan.

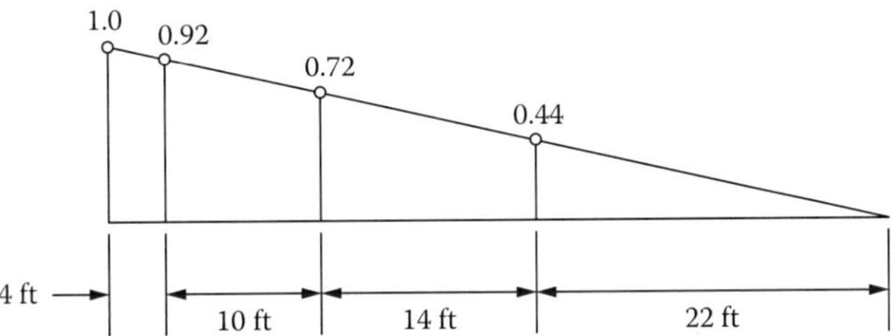

FIGURE 2.3b
Influence lines for shear at support.

$$N_L = \frac{w}{12\frac{ft}{lane}} = \frac{44.5\ ft}{12\ ft}$$

$$N_L = 3.7\ (3\ lanes)$$

Find the truck load moments and shears.

Design truck load (HS-20) for moment at midspan (Figure 2.3a).
Design tandem load for moment at midspan (Figure 2.3b).
The design truck load (HS-20) is shown in Figure 2.4. Figures 2.5 and 2.6 show the design tandem load position for moment at midspan and the design truck load (HS-20) position for shear at support, respectively. The design tandem load position for shear at support is shown in Figure 2.7.

NOTE: The maximum design truck or design tandem moment is generally used with the HL-93 center axle at midspan.

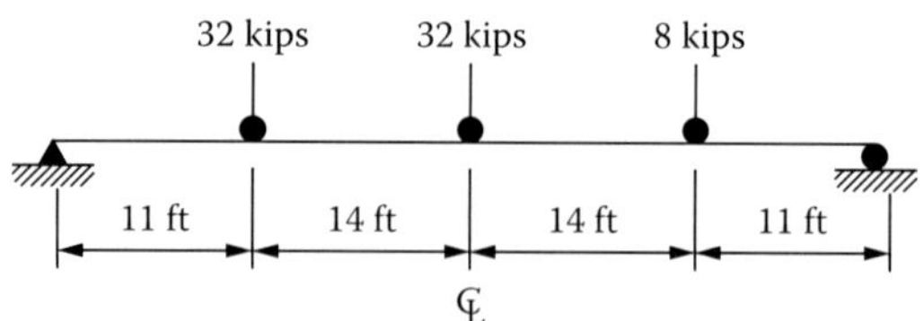

FIGURE 2.4
Design truck (HS-20) position for moment at midspan.

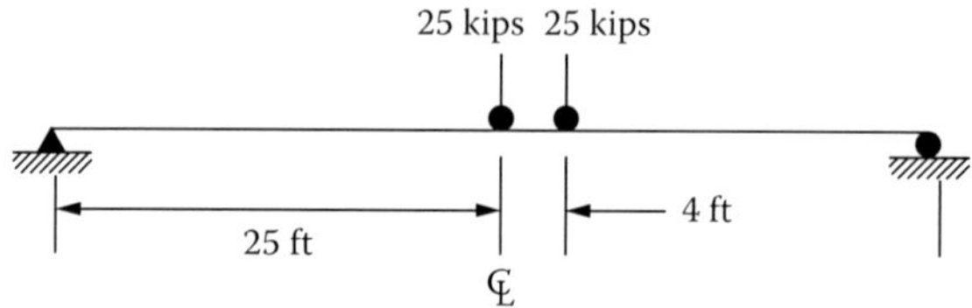

FIGURE 2.5
Design tandem load position for moment at midspan.

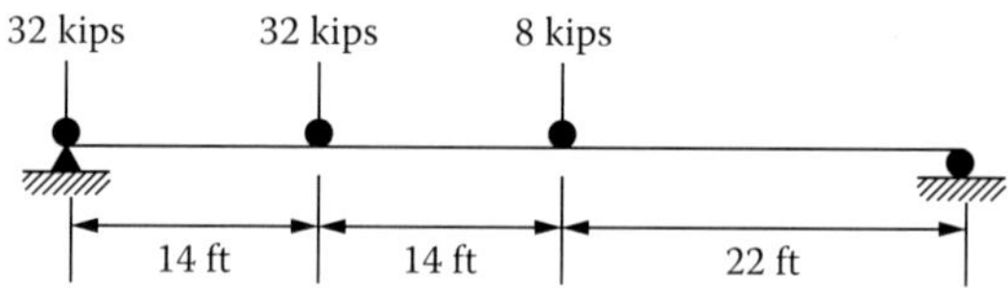

FIGURE 2.6
Design truck (HS-20) position for shear at support.

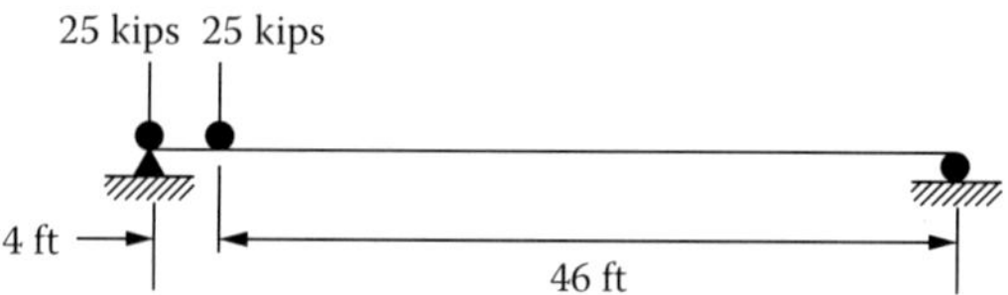

FIGURE 2.7
Design tandem load position for shear at support.

The truck load moment per lane is

$$M_{tr} = 32 \text{ kips } (12.5 \text{ ft} + 5.5 \text{ ft}) + 8 \text{ kips } (5.5 \text{ ft})$$

$$= 620 \text{ kip-ft}$$

The tandem load moment per lane is

$$M_{tandem} = 25 \text{ kips } (12.5 \text{ ft} + 10.5 \text{ ft})$$

$$= 575 \text{ kip-ft}$$

The truck load shear per lane is

$$V_{tr} = 32 \text{ kips } (1 + 0.72) + 8 \text{ kips } (0.44)$$

$$= 58.6 \text{ kips}$$

The tandem shear per lane is

$$V_{tandem} = 25 \text{ kips } (1 + 0.92)$$

$$= 48 \text{ kips}$$

The lane load per lane is

$$M_{ln} = \frac{wL^2}{8} = \frac{\left(0.64 \frac{\text{kips}}{\text{ft}}\right)(50 \text{ ft})^2}{8} = 200 \text{ kip-ft}$$

The lane load shear per lane is

$$V_{ln} = \frac{wL}{2} = \frac{\left(0.64 \frac{\text{kips}}{\text{ft}}\right)(50 \text{ ft})}{2} = 16 \text{ kips}$$

Find the live load distribution factors for moments, DFM.

For cast-in-place concrete T-beam, the deck type is (e).

A Tbl. 4.6.2.2.1-1

For interior beams,

A Tbl. 4.6.2.2.2; A Art. 4.6.2.2.2b; Tbl. 4.6.2.2.2b-1; Appendix A*

K_g longitudinal stiffness		in^4
L span length of beams		50 ft
N_b number of beams		5
S spacing of beams		10 ft
t_s slab thickness		9 in
$3.5 \text{ ft} \le S \le 16 \text{ ft}$	$S = 10$ ft	*OK*
$4.5 \text{ in} \le t_s \le 12 \text{ in}$	$t_s = 9$ in	*OK*
$20 \text{ ft} \le L \le 240 \text{ ft}$	$L = 50$ ft	*OK*
$N_b \ge 4$	$N_b = 5$	*OK*

* Refer to the Appendices at the end of the book.

The modular ratio between beam and deck material, n, is 1.0.

A Eq. 4.6.2.2.1-2

A	area of beam or T-beam	1710 in^2
L	span length	50 ft
t_s	slab thickness	9 in
n	modular ratio between beam and deck material	1

The moment of inertia of the basic beam (portion of beam below deck) is

$$I = \frac{(18\ \text{in})(35\ \text{in})^3}{12} = 64{,}312.5\ \text{in}^4$$

The distance between the centers of gravity of the basic beam and deck is

$$e_g = 17.5\ \text{in} + 4.5\ \text{in} = 22\ \text{in}$$

The longitudinal stiffness parameter is

A Eq. 4.6.2.2.1-1

$$K_g = n\,[I + (A)(e_g^{\,2})] = 1[64312.5\ \text{in}^4 + (1{,}710\ \text{in}^2)(22\ \text{in})^2]$$

$$K_g = 891{,}953.0\ \text{in}^4$$

A simplified value may be considered.

A Tbl. 4.6.2.2.1-2

$$\left[\frac{K_g}{12Lt_s^3}\right]^{0.1} = 1.05$$

Multiple presence factors, m, shall not be applied in conjunction with approximate load distribution factors specified in Art. 4.6.2.2 and 4.6.2.3, except where the lever rule is used.

A Art. 3.6.1.1.2

The distribution factor for moment for interior beams with one design lane, where *si* is the single lane loaded in interior beams is found as follows.

A Tbl. 4.6.2.2.2b-1 or Appendix A

The multiple presence factor, m, is applicable only when the lever rule is used for the distribution factors. Therefore, m = 1.0 where DFM is the distribution factor for moment.

$$\text{DFM}_{\text{si}} = \text{m}\left[0.06 + \left(\frac{\text{S}}{14}\right)^{0.4}\left(\frac{\text{S}}{\text{L}}\right)^{0.3}\left(\frac{\text{K}_\text{g}}{12\text{Lt}_\text{s}^{\ 3}}\right)^{0.1}\right]$$

$$= (1.0)\left[0.06 + \left(\frac{10\text{ ft}}{14}\right)^{0.4}\left(\frac{10\text{ ft}}{50\text{ ft}}\right)^{0.3}(1.05)\right]$$

$$= 0.629\text{ lane/girder}$$

The distribution factor for moment for interior beams with two or more design lanes loaded, where *mi* is the multiple lanes loaded in interior beams is

A Tbl. 4.6.2.2.2b-1 or Appendix A

$$\text{DFM}_{\text{mi}} = \left[0.075 + \left(\frac{\text{S}}{9.5}\right)^{0.6}\left(\frac{\text{S}}{\text{L}}\right)^{0.2}\left(\frac{\text{K}_\text{g}}{12\text{ Lt}_\text{s}^3}\right)^{0.1}\right]$$

$$= \left[0.075 + \left(\frac{10\text{ ft}}{9.5}\right)^{0.6}\left(\frac{10\text{ ft}}{50\text{ ft}}\right)^{0.2}(1.05)\right]$$

$$= 0.859\text{ lane/girder}\left[\text{governs for interior beams}\right]$$

For the distribution of moment for exterior beams with one design lane, use the lever rule. See Figure 2.8.

A Art. 3.6.1.3.1, A Art. 4.6.2.2.2d; Tbl. 4.6.2.2.2d-1 or Appendix B

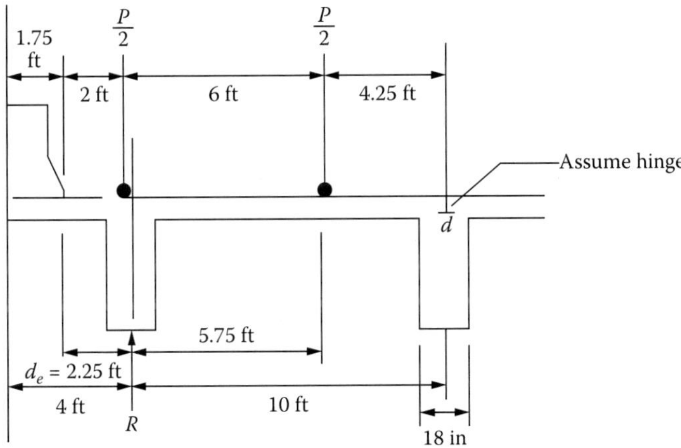

FIGURE 2.8
Lever rule for determination of distribution factor for moment in exterior beam, one lane loaded.

Σ Moment at "a" = 0

$$0 = -R(10\ \text{ft}) + \frac{P}{2}(10.25\ \text{ft}) + \frac{P}{2}(4.25\ \text{ft})$$

$$R = 0.725\ P$$

The multiple presence factor for one design lane loaded, m, is 1.20.

A Tbl. 3.6.1.1.2-1

The distribution factor for moment for exterior beams for one design lane loaded is, where $_{se}$ is the designation for single lane loaded in the extreme beam,

$$DFM_{se} = m[0.725] = 1.20\ [0.725]$$

$$= 0.87\ \text{lane/girder [governs for exterior beam]}$$

The exterior web of the exterior beam to the interior edge of the curb, d_e, is 2.25 ft, which is OK for the –1.0 ft ≤ d_e ≤ 5.5 ft range.

A Tbl. 4.6.2.2.2d-1 or Appendix B

The distribution factor for the exterior beam is

$$(e)(g_{interior}) = (e)(DFM_{mi})$$

$$e = 0.77 + \frac{d_e}{9.1}$$

$$e = 0.77 + \frac{2.25}{9.1}$$

$$e = 1.017$$

Use e = 1.0.

The distribution factor for moment for exterior beams with two or more design lanes loaded, where me is the designation for multiple lanes loaded in the exterior beam, is

A Tbl. 3.6.1.1.2-1

$$DFM_{me} = (m)(e)(g_{interior}) = (m)(e)(DFM_{mi}) = (0.85)(1.0)(0.859)$$

$$= 0.730\ \text{lane/girder}$$

Find the distributed live load moments.

The governing distribution factors are:

Interior beam DFM_{mi} = 0.859 lane/girder
Exterior beam DFM_{se} = 0.870 lane/girder

IM = 33%

A Tbl. 3.6.2.1-1

The unfactored live load moment per beam for interior beams due to truck load is

$$M_{TL} = M_{tr}(DFM)(1 + IM) = (620 \text{ ft-kips})(0.859)(1+0.33)$$

$$= 708.33 \text{ ft-kips}$$

The unfactored live load moment per beam for interior beams due to lane load is

$$M_{LN} = M_{ln}\,(DFM) = (200 \text{ ft-kips})(0.859)$$

$$= 171.8 \text{ ft-kips}$$

The unfactored live load moment per beam for exterior beams due to truck load is

$$M_{TL} = M_{tr}(DFM)(1 + IM) = (620 \text{ ft-kips})(0.87)(1 + 0.33)$$

$$= 717.40 \text{ ft-kips}$$

The unfactored live load moment per beam for exterior beams due to lane load is

$$M_{LN} = M_{ln}\,(DFM) = (200 \text{ ft-kips})(0.87)$$

$$= 174.0 \text{ ft-kips}$$

Find the distribution factors for shears, DFV.

A Art. 4.6.2.2.3

For a cast-in-place concrete T-beam, the deck type is (e).

A Tbl. 4.6.2.2.1-1

The distribution factor for shear for interior beams with one design lane loaded is

A Tbl. 4.6.2.2.3a-1 or Appendix C

$$DFV_{si} = \left[0.36 + \frac{S}{25}\right] = \left[0.36 + \frac{10\text{ ft}}{25}\right]$$

$$= 0.76\text{ lanes}$$

The distribution factor for shear for interior beams with two or more design lanes loaded is

$$DFV_{mi} = \left[0.2 + \frac{S}{12} - \left(\frac{S}{35}\right)^{2.0}\right] = \left[0.2 + \frac{10\text{ ft}}{12} - \left(\frac{10\text{ ft}}{35}\right)^{2.0}\right]$$

$DFV_{mi} = 0.95$ lane/girder [controls for interior beams]

Using the lever rule for moment, the distribution factor for shear for exterior beams with one design lane loaded is

A Tbl. 4.6.2.2.3b-1

$DFV_{se} = DFM_{se} = 0.87$ lane/girder [controls for exterior beams]

The distribution factor for shear for exterior beams with two or more design lanes loaded is

$$g = (e)g_{interior} = (e)(DFV_{mi})$$

$$e = 0.6 + \frac{d_e}{10} = 0.6 + \frac{2.25\text{ ft}}{10} = 0.825$$

$DFV_{me} = mg = (m)(e)(g_{interior})$, where $g_{interior} = DFV_{mi}$

A Tbl. 3.6.1.1.2-1

$$= (1.0)(0.825)(0.95)$$

$$= 0.784\text{ lane/girder}$$

Find the distributed live load shears for Strength I.

The governing distribution factors are:

Interior beam $DFV_{mi} = 0.95$ lane/girder
Exterior beam $DFV_{se} = 0.87$ lane/girder

The unfactored live load shears per beam due to truck load for interior beams is

$$V_{TL} = V_{tr}(DFV)(1 + IM)$$

$$= (58.6 \text{ kips})(0.95)(1.33)$$

$$= 74.04 \text{ kips}$$

The unfactored live load shear per beam due to lane load for interior beams is

$$V_{LN} = V_{ln}(DFV)$$

$$= (16.0 \text{ kips})(0.95)$$

$$= 15.2 \text{ kips}$$

The unfactored live load shears per beam for exterior beams are

$$V_{TL} = V_{tr}(DFV)(1 + IM)$$

$$= (58.6 \text{ kips})(0.87)(1.33)$$

$$= 67.80 \text{ kips}$$

$$V_{LN} = V_{ln}(DFV)$$

$$= (16.0 \text{ kips})(0.87)$$

$$= 13.92 \text{ kips}$$

Find the dead load force effects.

The self-weights of the T-beam, the deck, and the curb and parapet for interior beams are represented by the variable DC.

$$DC_{T\text{-beam}} = \left(1710 \text{ in}^2\right)\left(\frac{1 \text{ ft}^2}{144 \text{ in}^2}\right)\left(0.15 \frac{\text{kips}}{\text{ft}^3}\right)$$

$$= 1.78 \frac{\text{kips}}{\text{ft}}$$

$$DC_{C\&P} = 2\left(3.37 \text{ ft}^2\right)\left(0.15 \frac{\text{kips}}{\text{ft}^3}\right)\left(\frac{1}{5 \text{ beams}}\right)$$

$$= 0.202 \frac{\text{kips}}{\text{ft}}$$

$$w_{DC} = DC_{T\text{-beam}} + DC_{C\&P}$$

$$= 1.78\,\frac{\text{kips}}{\text{ft}} + 0.202\,\frac{\text{kips}}{\text{ft}}$$

$$= 1.98\,\frac{\text{kips}}{\text{ft}}$$

The corresponding shear is

$$V_{DC} = \frac{wL}{2} = \frac{\left(1.98\,\frac{\text{kips}}{\text{ft}}\right)(50\text{ ft})}{2}$$

$$= 49.5\text{ kips}$$

The corresponding moment is

$$M_{DC} = \frac{wL^2}{8} = \frac{\left(1.98\,\frac{\text{kips}}{\text{ft}}\right)(50\text{ ft})^2}{8}$$

$$= 618.75\text{ kip-ft}$$

The future wearing surface load, DW, per beam is

$$w_{DW} = \left(0.03\,\frac{\text{kips}}{\text{ft}^2}\right)\left(10\,\frac{\text{ft}}{\text{beam}}\right)$$

$$= 0.3\,\frac{\text{kips}}{\text{ft}}$$

The corresponding shear is

$$V_{DW} = \frac{w_{DW}L}{2} = \frac{\left(0.3\,\frac{\text{kips}}{\text{ft}}\right)(50\text{ ft})}{2}$$

$$= 7.5\text{ kips}$$

TABLE 2.1

Distributed Live Load and Dead Load Effects for Interior Beam for Reinforced Concrete T-Beam Bridge

	Moment M (ft-kips)	Shear V (kips)
DC	618.75	49.5
DW	93.75	7.5
TL	708.33	74.04
LN	171.8	15.2

The corresponding moment is

$$M_{DW} = \frac{w_{DW}L^2}{8} = \frac{\left(0.3\,\frac{\text{kips}}{\text{ft}}\right)\left(50\text{ ft}\right)^2}{8}$$

$$= 93.75\,\text{kip-ft}$$

The unfactored interior beam moments and shears due to the dead loads plus the live loads are given in Table 2.1. Note that dead loads consist of the exterior T-beam stem, deck slab, curb/parapet, and wearing surface.

For exterior girders, the deck slab load is,

$$w_s = (9\text{ in})\left(\frac{1\text{ ft}}{12\text{ in}}\right)\left(0.15\,\frac{\text{kips}}{\text{ft}^3}\right)$$

$$= 0.113\,\frac{\text{kips}}{\text{ft}^2}$$

For exterior girders, the wearing surface load is given as

$$w_{DW} = 0.03\,\frac{\text{kips}}{\text{ft}^2}$$

The total load is

$$w = w_S + w_{DW}$$

$$= 0.113\,\frac{\text{kips}}{\text{ft}^2} + 0.03\,\frac{\text{kips}}{\text{ft}^2}$$

$$= 0.143\,\frac{\text{kips}}{\text{ft}^2}$$

See Figure 2.9.

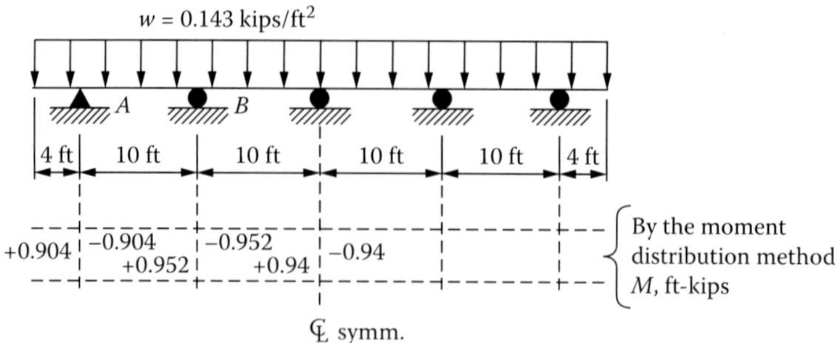

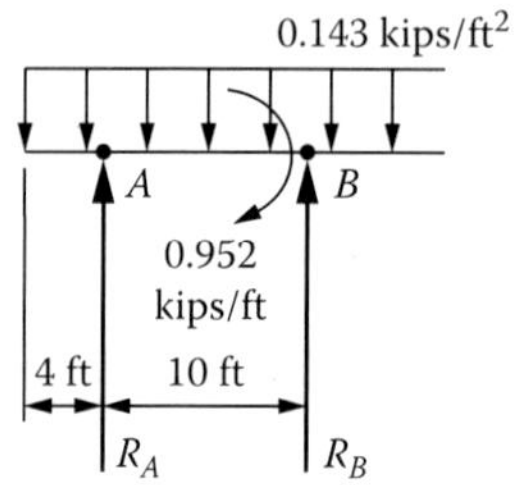

FIGURE 2.9
Moment distribution for deck slab and wearing surface loads.

Σ moments about B = 0

$$0 = -0.952 \text{ ft-kips} - R_A(10 \text{ ft}) + (0.143 \text{ kips/ft}^2)(14 \text{ ft})(7 \text{ ft}).$$

R_A = 1.31 kips per foot of exterior beam due to the deck slab and wearing surface dead loads

The load for the curb and parapet for exterior girders is,

$$w_{C\&P} = \left(3.37 \text{ ft}^2\right)\left(0.15 \frac{\text{kips}}{\text{ft}^3}\right)$$

$$= 0.51 \frac{\text{kips}}{\text{ft}}$$

See Figure 2.10.

Σ moments about B = 0

$$0 = 0.476 \text{ ft-kips} - R_A(10 \text{ ft}) + (0.51 \text{ kips})(13.34 \text{ ft})$$

R_A = 0.728 kips per foot of exterior beam due to the curb and parapet dead loads

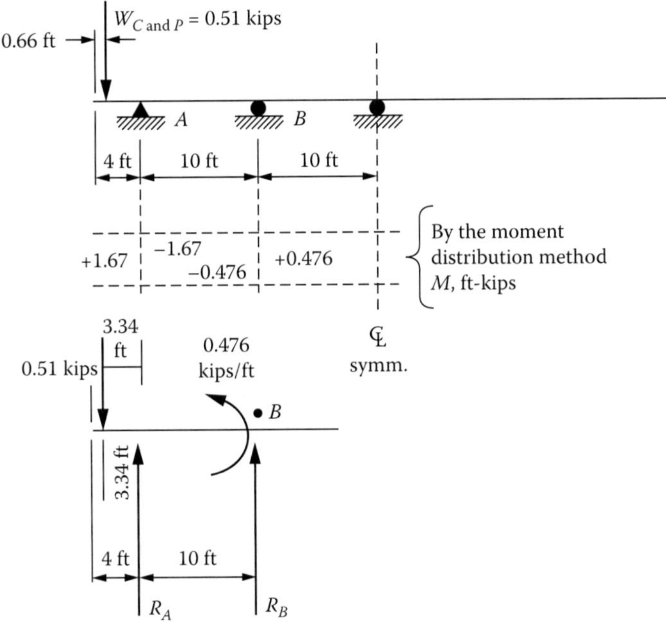

FIGURE 2.10
Moment distribution for curb and parapet loads for exterior girder.

The reactions at exterior beam due to the dead loads are as follows

$$\text{DC deck slab} = \left(1.31\,\frac{\text{kips}}{\text{ft}}\right)\left(\frac{0.113\,\dfrac{\text{kips}}{\text{ft}^2}}{0.143\,\dfrac{\text{kips}}{\text{ft}^2}}\right) = 1.04\,\frac{\text{kips}}{\text{ft}}$$

Curb and parapet overhang

$$w_{C\&P} = 0.728\,\frac{\text{kips}}{\text{ft}}$$

$$\text{girder stem} = \left(630\text{in}^2\right)\left(\frac{1\ \text{ft}^2}{144\ \text{in}^2}\right)\left(0.15\,\frac{\text{kips}}{\text{ft}^3}\right) = 0.656\,\frac{\text{kips}}{\text{ft}}$$

$$w_{DC} = 1.04\,\frac{\text{kips}}{\text{ft}} + 0.728\,\frac{\text{kips}}{\text{ft}} + 0.656\,\frac{\text{kips}}{\text{ft}} = 2.42\,\frac{\text{kips}}{\text{ft}}$$

Future wearing surface

$$w_{DW} = \left(1.31\frac{\text{kips}}{\text{ft}}\right)\left(\frac{0.03\dfrac{\text{kips}}{\text{ft}^2}}{0.143\dfrac{\text{kips}}{\text{ft}^2}}\right) = 0.27\frac{\text{kips}}{\text{ft}}$$

The unfactored exterior girder moments and shears due to the dead loads plus the live loads are as follows.

$$M_{DC} = \frac{w_{DC}L^2}{8} = \frac{\left(2.42\dfrac{\text{kips}}{\text{ft}}\right)(50\text{ ft})^2}{8} = 756.3\text{ kip-ft}$$

$$V_{DC} = \frac{w_{DC}L}{2} = \frac{\left(2.42\dfrac{\text{kips}}{\text{ft}}\right)(50\text{ ft})}{2} = 60.5\text{ kips}$$

$$M_{DW} = \frac{w_{DW}L^2}{8} = \frac{\left(0.27\dfrac{\text{kips}}{\text{ft}}\right)(50\text{ ft})^2}{8} = 84.4\text{ kip-ft}$$

$$V_{DW} = \frac{w_{DW}L}{2} = \frac{\left(0.27\dfrac{\text{kips}}{\text{ft}}\right)(50\text{ ft})}{2} = 6.75\text{ kips}$$

The unfactored beam moments and shears due to the dead loads plus the live loads are given in Table 2.2.

TABLE 2.2

Unfactored Beam Moments and Shears Due to Dead Loads and Live Loads for Reinforced Concrete T-Beam Bridge

	Interior Beam		Exterior Beam	
	Moment M (ft-kips)	**Shear V (kips)**	**Moment M (ft-kips)**	**Shear V (kips)**
DC	618.75	49.5	756.3	60.5
DW	93.75	7.5	84.4	6.75
TL	708.33	74.04	717.4	67.8
LN	171.8	15.2	174.0	13.92

Find the factored moments and shears for Strength I.

A Tbls. 3.4.1-1; 3.4.1-2]

$$Q = 1.25\ DC + 1.5\ DW + 1.75(TL + LN)$$

For interior girders, the unfactored moment is

$$M_u = 1.25\ M_{DC} + 1.5\ M_{DW} + 1.75\ (M_{TL} + M_{LN})$$

$$= 1.25\ (618.75\ \text{kip-ft}) + 1.5\ (93.75\ \text{kip-ft}) + 1.75\ (708.33\ \text{kip-ft} + 171.8\ \text{kip-ft})$$

$$= 2454.29\ \text{kip-ft}$$

For interior girders, the factored shear is

$$V_u = 1.25\ V_{DC} + 1.5\ V_{DW} + 1.75\ (V_{TL} + V_{LN})$$

$$= 1.25\ (49.5\ \text{kips}) + 1.5\ (7.5\ \text{kips}) + 1.75\ (74.04\ \text{kips} + 15.2\ \text{kips})$$

$$= 229.2\ \text{kips [controls]}$$

For exterior girders, the factored moment is

$$M_u = 1.25\ M_{DC} + 1.5\ M_{DW} + 1.75\ (M_{TL} + M_{LN})$$

$$= 1.25\ (756.3\ \text{kip-ft}) + 1.5\ (84.4\ \text{kip-ft}) + 1.75\ (717.4\ \text{kip-ft} + 174.0\ \text{kip-ft})$$

$$= 2631.9\ \text{kip-ft [controls]}$$

For exterior girders, the factored shear is

$$V_u = 1.25\ V_{DC} + 1.5\ V_{DW} + 1.75\ (V_{TL} + V_{LN})$$

$$= 1.25\ (60.5\ \text{kips}) + 1.5\ (6.75\ \text{kips}) + 1.75\ (67.8\ \text{kips} + 13.92\ \text{kips})$$

$$= 228.8\ \text{kips}$$

Find the design flexural reinforcements, neglecting compression reinforcement, and note the exterior girder moment and interior girder shear control.

$$M_u = 2631.9\ \text{kip-ft}$$

$$V_u = 229.2\ \text{kips}$$

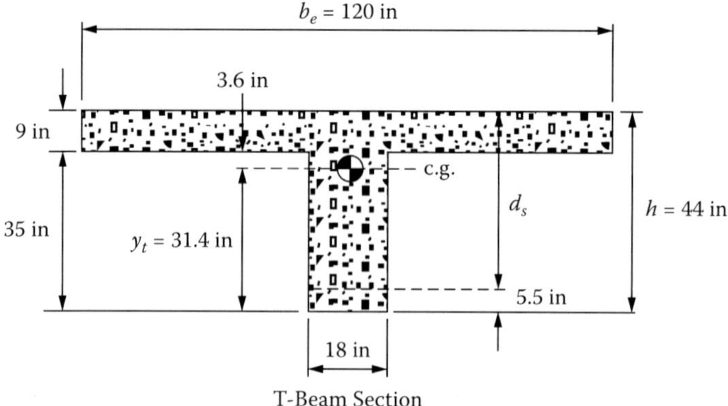

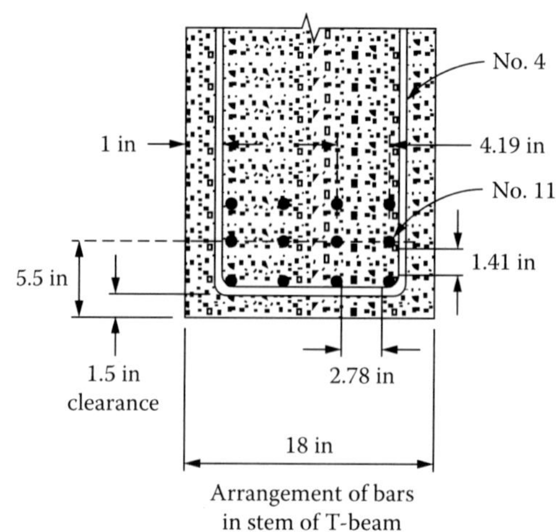

FIGURE 2.11
T-beam section and reinforcement in T-beam stem.

For stem reinforcement, try 12 no. 11 bars. See Figure 2.11.
Web thickness, b_w, is,

A Art. 5.14.1.5.1c

$$b_w = 2(1\text{ in}) + 2(0.5\text{ in}) + 4\,d_b + 3(1.5\,d_b)$$

$$= 2\text{ in} + 1\text{ in} + 4(1.4\text{ in}) + 3(1.5 \times 1.4\text{ in})$$

$$= 14.9\text{ in}$$

To give a little extra room, use b_w = 18 in

The distance from extreme compression fiber to the centroid of tensile reinforcement is

$$d_s = 44 \text{ in} - 5.5 \text{ in}$$

$$= 38.5 \text{ in}$$

$$d_b = 1.41 \text{ in [no. 11 bar]}$$

For 12 no. 11 bars, $A_s = 18.72 \text{ in}^2$.

The minimum clearance between bars in a layer must not be less than

A Art. 5.10.3.1.1

$$1.5\ d_b = 1.5\ (1.41 \text{ in}) = 2.1 \text{ in}$$

$$2.1 \text{ in} < \text{provided} = 2.78 \text{ in [OK]}$$

The clear distance between layers shall not be less than 1 in or d_b.

A Art. 5.10.3.1.3

$$d_b = 1.41 \text{ in [OK]}$$

Concrete cover for epoxy-coated main reinforcing bars is 1.0 in, provided = 1.5 in [OK]

A Art. 5.12.3

$$\Phi = 0.90$$

A Art. 5.5.4.2

$$A_s = 18.75 \text{ in}^2$$

$$f_y = 60.0 \text{ ksi}$$

$$d_s = 38.5 \text{ in}$$

The thickness of the deck slab, t_s, is 9 in and the width of compression face, b, is 120 in.

The factor for concrete strength β_1 is

A Art. 5.7.2.2

$$\beta_1 = 0.85 - \left(\frac{f'_c - 4000 \dfrac{\text{kips}}{\text{in}^2}}{1000 \dfrac{\text{kips}}{\text{in}^2}} \right)(0.05)$$

$$= 0.85 - \left(\frac{4500 \dfrac{\text{kips}}{\text{in}^2} - 4000 \dfrac{\text{kips}}{\text{in}^2}}{1000 \dfrac{\text{kips}}{\text{in}^2}} \right)(0.05)$$

$$= 0.825$$

The distance from the extreme compression fiber to the neutral axis, c, is

A Eq. 5.7.3.1.1-4

$$c = \frac{A_s f_y}{0.85\, f'_c\, \beta_1\, b}$$

$$= \frac{(18.75\ \text{in}^2)\left(60 \dfrac{\text{kips}}{\text{in}^2}\right)}{0.85\left(4.5 \dfrac{\text{kips}}{\text{in}^2}\right)(0.825)(120\ \text{in})}$$

$$c = 2.97\ \text{in}$$

$$a = c\, \beta_1$$

$$= (2.97\ \text{in})(0.825)$$

$$= 2.45\ \text{in} < t_s = 9\ \text{in}\ [\text{OK}]$$

The nominal resisting moment, M_n, is

A Art. 5.7.3.2.2

$$M_n = A_s f_y \left(d_s - \frac{a}{2} \right)$$

$$= (18.75\ \text{in}^2)\left(60 \frac{\text{kips}}{\text{in}^2}\right)\left(38.5\ \text{in} - \frac{2.45\ \text{in}}{2}\right)\left(\frac{1\ \text{ft}}{12\ \text{in}}\right)$$

$$= 3488.9\ \text{in}^2$$

The factored resisting moment, M_r, is

A Art. 5.7.3.2.1

$$M_r = \Phi M_n$$

$$= 0.90(3488.9 \text{ ft-kips})$$

$$= 3140.0 \text{ ft-kips} > M_u = 2631.9 \text{ ft-kips [OK]}$$

Check the reinforcement requirements.

A Art. 5.7.3.3.1; 5.7.3.3.2

This provision for finding the maximum reinforcement was deleted from AASHTO in 2005. Therefore, check for the minimum reinforcement. The minimum reinforcement requirement is satisfied if M_r is at least equal to the lesser of:

The minimum reinforcement requirement is satisfied if ΦM_n (= M_r) is at least equal to the lesser of:

- $M_r \geq 1.2\, M_{cr}$
- $M_r \geq 1.33$ times the factored moment required by the applicable strength load combination specified in AASHTO Table 3.4.1-1

where:
M_r = factored flexural resistance
M_{cr} = cracking moment
f_r = modulus of rupture

$$M_{cr} = S_c f_r$$

A Eq. 5.7.3.3.2-1

The section modulus is

$$S_c = \frac{I_g}{y_b}$$

The modulus of rupture of concrete is

A Art. 5.4.2.6

$$f_r = 0.37\sqrt{f'_c}$$

$$= 0.37\sqrt{4.5\frac{\text{kips}}{\text{in}^2}}$$

$$= 0.7849\frac{\text{kips}}{\text{in}^2}$$

$$M_{cr} = \left(\frac{I_g}{y_t}\right)(f_r)$$

$$= \left(\frac{264{,}183.6\ \text{in}^4}{31.4\ \text{in}}\right)\left(0.7849\frac{\text{kips}}{\text{in}^2}\right)\left(\frac{1\ \text{ft}}{12\ \text{in}}\right)$$

$$= 550.3\ \text{kip-ft}$$

$$1.2\ M_{cr} = (1.2)(550.3\ \text{kip-ft})$$

$$= 660.4\ \text{kip-ft [controls]}$$

$$1.33\ M_u = (1.33)(2631.9\ \text{kip-ft})$$

$$= 3500.4\ \text{kip-ft}$$

$$M_r = \Phi M_n = 0.9(3488.9\ \text{kip-ft}) = 3140.0\ \text{kip-ft} > 1.2\ M_{cr} = 660.4\ \text{kip-ft [OK]}$$

Design for shear.

The effective shear depth, d_v, taken as the distance between the resultants of the tensile and compressive forces due to flexure

A Art. 5.8.2.9

$$d_v = d_s - \frac{a}{2} = 38.5\ \text{in} - \frac{2.45\ \text{in}}{2} = 37.3\ \text{in}$$

where:

d_s = distance from the extreme compression fiber to the centroid of tensile reinforcement

The effective shear depth, d_v, need not be less than the greater of:

A Art. 5.8.2.9

- $0.9\ d_e = 0.9(38.5\ \text{in}) = 34.6\ \text{in} < 37.3\ \text{in}$
- $0.72\ h = 0.72(44\ \text{in}) = 31.7\ \text{in} < 37.3\ \text{in}$

The critical section for shear is taken as d_v from the internal face of support. The distance from the center of bearing support, x, is calculated as

A Art. 5.8.3.2

$$x = 37.3\ \text{in} + 5.7\ \text{in} = 43\ \text{in}\left(\frac{1\ \text{ft}}{12\ \text{in}}\right) = 3.58\ \text{ft}$$

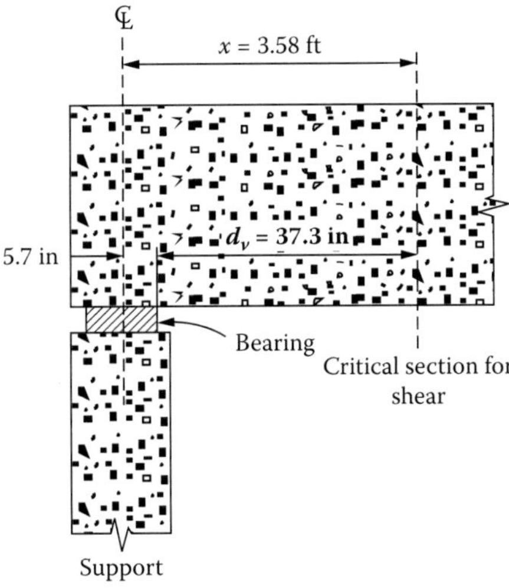

FIGURE 2.12
Critical shear section at support.

See Figure 2.12.

Find the unfactored shear forces and concurrent moments due to dead load.

Note that the exterior girder controls for moment.

The shear at the critical section for shear due to the DC load [interior girder controls] is

$$w_{DC} = 1.98 \text{ kips/ft}$$

$$V_{DC} = \left(\frac{1.98 \frac{\text{kips}}{\text{ft}} (50 \text{ ft})}{2} \right) - \left(1.98 \frac{\text{kips}}{\text{ft}} \right) (3.58 \text{ ft})$$

$$V_{DC} = 42.2 \text{ kips}$$

The concurrent moment at the critical section for shear for the interior girder due to the DC load is

$$M_{DC} = \left(\frac{50 \text{ ft}}{2} \right) \left(1.98 \frac{\text{kips}}{\text{ft}} \right) (3.58 \text{ ft}) - \left(1.98 \frac{\text{kips}}{\text{ft}} \right) (3.58 \text{ ft})^2 \left(\frac{1}{2} \right)$$

$$= 189.9 \text{ kip-ft}$$

The shear at the critical section for shear due to the DW load is

$$w_{DW} = 0.30 \text{ kips/ft}$$

$$V_{DW} = \frac{\left(0.30\dfrac{\text{kips}}{\text{ft}}\right)(50 \text{ ft})}{2} - \left(0.30\dfrac{\text{kips}}{\text{ft}}\right)(3.58 \text{ ft})$$

$$V_{DW} = 6.43 \text{ kips}$$

The concurrent moment at the critical section for shear due to the DW load is

$$M_{DW} = \frac{(50 \text{ ft})\left(0.30\dfrac{\text{kips}}{\text{ft}}\right)}{2}(3.58 \text{ ft}) - \left(0.30\dfrac{\text{kips}}{\text{ft}}\right)(3.58 \text{ ft})^2\left(\frac{1}{2}\right)$$

$$M_{DW} = 24.93 \text{ kip-ft}$$

Find the shear at the critical section for shear for the interior girder due to the design truck (HS-20).

MBE-2 App Tbl. H6B*

Let x = L − x; multiply by 2 for two wheel lines.

$$V = \frac{36 \text{ kips}(x - 9.33)}{L}(2)$$

$$V_{tr,\,HS\text{-}20} = \frac{72 \text{ kips}(L - x - 9.33)}{L}$$

$$= \frac{72 \text{ kips}(50 \text{ ft} - 3.58 \text{ ft} - 9.33)}{50 \text{ ft}}$$

$$= 53.41 \text{ kips per lane}$$

$$V_{TL} = V_{tr}(DFV)(1 + IM)$$

$$= (53.71 \text{ kips})(0.95)(1.33)$$

$$= 67.48 \text{ kips per beam}$$

* MBE-2 refers to the *Manual for Bridge Evaluation* (second edition), 2011, American Association of State Highway and Transportation Officials (AASHTO)

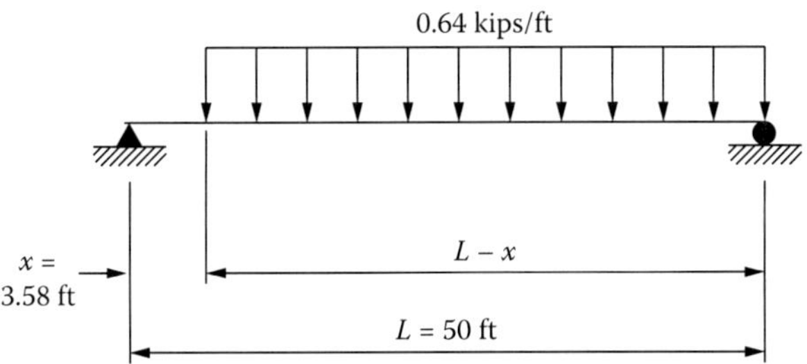

FIGURE 2.13
Lane load position for maximum shear at critical shear section.

Find the concurrent maximum moment per lane interior girder at the critical section for shear due to the design truck (HS-20).

MBE-2 App. Tbl. J6B

Let x = L − x; multiply by 2 for two wheel lines.

$$M = \frac{36\ \text{kips}(L-x)(x-9.33)}{L}(2)$$

$$M_{tr,\ HS\text{-}20} = \frac{72\ \text{kips}(x)(L-x-9.33)}{L}$$

$$= \frac{72\ \text{kips}(3.58\ \text{ft})(50\ \text{ft}-3.58\ \text{ft}-9.33)}{50\ \text{ft}}$$

$$= 191.2\ \text{ft-kips per lane}$$

$$M_{TL} = M_{tr}(\text{DFM})(1 + \text{IM})$$

$$= (191.2\ \text{ft-kips})(0.822)(1.33)$$

$$= 209.03\ \text{ft-kips per beam}$$

Find the shear at the critical section for interior girder due to lane load. See Figure 2.13.

$$V_{ln} = \frac{0.32(L-x)^2}{L}$$

$$= \frac{0.32(50\ \text{ft}-3.58\ \text{ft})^2}{50\ \text{ft}}$$

$$= 13.79\ \text{kips per lane}$$

$$V_{LN} = V_{ln}(DFV) = (13.79 \text{ kips})(0.95)$$

$$= 13.1 \text{ kips per beam}$$

Find the concurrent moment at the critical section for shear for interior girder due to lane load, where

$$R = \frac{0.32(L-x)^2}{L}$$

$$M_{ln} = Rx$$

$$= \frac{0.32(L-x)^2(x)}{L}$$

$$= \frac{0.32(50 \text{ ft} - 3.58 \text{ ft})^2(3.58 \text{ ft})}{50 \text{ ft}}$$

$$= 49.37 \text{ ft-kips per beam}$$

$$M_{LN} = M_{ln}(DFM) = (49.37 \text{ ft-kips})(0.822)$$

$$= 40.58 \text{ ft-kips per beam}$$

The factored shear at the critical section is,

A Tbls. 3.4.1-1; 3.4.1-2

$$V_u = 1.25\ V_{DC} + 1.5\ V_{DW} + 1.75(V_{TL} + V_{LN})$$

$$= 1.25(42.2 \text{ kips}) + 1.5(6.43 \text{ kips}) + 1.75(67.48 \text{ kips} + 13.1 \text{ kips})$$

$$V_u = 203.4 \text{ kips}$$

The factored moment at the critical section is

$$M_u = 1.25\ M_{DC} + 1.5\ M_{DW} + 1.75(M_{TL} + M_{LN})$$

$$= 1.25(189.9 \text{ kip-ft}) + 1.5(24.93 \text{ kip-ft}) + 1.75(209.03 \text{ kip-ft} + 40.58 \text{ kip-ft})$$

$$M_u = 711.59 \text{ kip-ft}$$

Find the shear stress on the concrete.

A Eq. 5.8.2.9-1

The effective web width, b_v, is 18 in.
The resistance factor for shear specified for reinforced concrete is 0.9.

A Art. 5.5.4.2

The distance from extreme compression fiber to the centroid of tensile reinforcement, d_s, is 38.5 in (d_s is equivalent to d_e).

A Art. 5.8.2.9

NOTE: d_e is also the distance from the centerline of the exterior web of the exterior beam to the interior edge of curb or traffic barrier (Art. 4.6.2.2.1).

The factored shear stress on the concrete at d_v = 37.3 in is,

A Eqs. 5.8.2.9-1, 5.5.4.2.1

$$v_u = \frac{V_u}{\Phi b_v d_v} = \frac{203.4 \text{ kips}}{(0.9)(18 \text{ in})(37.3 \text{ in})} = 0.34 \frac{\text{kips}}{\text{in}^2}$$

Find the tensile strain in the transverse reinforcement for sections where the transverse reinforcement is found using Eq. 5.8.3.4.2-1 and has the following characteristics.

A Art. 5.8.3.4.2

$$\theta = 30°$$

$$\cot\theta = 1.732$$

$$E_s = 29{,}000 \text{ kips/in}^2$$

$$A_s = 18.75 \text{ in}^2$$

$$d_v = 37.3 \text{ in}$$

$$\varepsilon_x = \frac{\dfrac{|M_u|}{d_v} + 0.5\, V_u \cot\theta}{2E_sA_s}$$

A App. B5 Eq. B5.2-1

$$= \frac{\left(\dfrac{711.59 \text{ ft-kips}}{37.3 \text{ in}}\right)\left(12\dfrac{\text{in}}{\text{ft}}\right) + 0.5(203.4 \text{ kips})(1.732)}{2\left(29000\dfrac{\text{kips}}{\text{in}^2}\right)(18.75 \text{ in}^2)}$$

$$= 0.00037 < 0.001 \text{ [OK]}$$

Transverse reinforcement shall be provided where:

A Eq. 5.8.2.4-1; Arts. 5.8.3.4.1

$$V_u \geq 0.5\ \Phi V_c$$

$\beta = 2$ and $\theta = 45°$ may be used.

The nominal shear resistance, V_n, shall be as the lesser of:

A Art. 5.8.3.3; Eq. 5.8.3.3-1; Eq. 5.8.3.3-2

$$V_n = V_c + V_s$$

$$V_n = 0.25\ f_c' b_v d_v$$

The nominal shear resistance by the concrete is

$$V_c = 0.0316\,\beta\sqrt{f_c'}\ b_v d_v$$

$$= 0.0316(2)\sqrt{4.5\frac{\text{kips}}{\text{in}^2}}\,(18\text{ in})(37.3\text{ in})$$

$$= 90.1\text{ kips}$$

Transverse reinforcement shall be provided if $V_u \geq 0.5\ \Phi V_c$

A Art. 5.8.2.4

$$0.5\ \Phi V_c = (0.5)(0.9)(90.1\text{ kips})$$

$$= 40.5\text{ kips} < V_u = 203.4\text{ kips}$$

Therefore, the transverse reinforcement is provided at the critical section ($x = 3.58$ ft) for shear.

The nominal shear resistance of a transversely reinforced section is $V_n = V_c + V_s$. V_s is the shear resistance by reinforcement.

A Eq. 5.8.3.3-1

The nominal shear resistance of the section is

A Eq. 5.8.3.3-2

$$V_n = 0.25\,f_c' b_v d_v = (0.25)\left(4.5\frac{\text{kips}}{\text{in}^2}\right)(18\text{ in})(37.3\text{ in})$$

$$= 755.3\text{ kips}$$

The factored shear resistance, V_r, is

A Eq. 5.8.2.1-2

$$V_r = \Phi V_n = (0.9)(755.3 \text{ kips})$$

$$= 679.8 \text{ kips} > V_u = 203.4 \text{ kips [OK]}$$

The factored shear force does not exceed the maximum factored shear resistance, therefore, the section size is good for shear.

Find the transverse reinforcement requirements.

The maximum shear resistance provided by shear reinforcement is

A Art. 5.8.3.3

$$V_s = V_n - V_c = 0.25 f_c' b_v d_v - 0.0316\beta\sqrt{f_c'}\, b_v d_v$$

$$= 755.3 \text{ kips} - 90.1 \text{ kips}$$

$$= 665.2 \text{ kips}$$

Shear required by shear reinforcement at the critical section is determined by letting the nominal shear resistance, V_n, equal to the factored shear forces, V_u, divided by Φ.

$$V_{s.\text{required}} = V_n - V_c = \frac{V_u}{\Phi} - V_c$$

$$= \frac{203.4 \text{ kips}}{0.9} - 90.1 \text{ kips}$$

$$= 136.0 \text{ kips} < 665.2 \text{ kips [OK]}$$

θ = angle of inclination of diagonal compressive stress = 45°

α = angle of inclination of transverse reinforcement to longitudinal axis = 90°

A Comm. 5.8.3.3, A Eq. 5.8.3.3-4

$$V_s = \frac{A_v f_y d_v (\cot\theta + \cot\alpha)\sin\alpha}{s}$$

$$= \frac{A_v f_y d_v \left(\cot 45° + \cot 90°\right)\left(\sin 90°\right)}{s} = \frac{A_v f_y d_v \left(1.0 + 0\right)\left(1.0\right)}{s}$$

Try two legs of no. 4 bar stirrups at the critical section, $A_v = 0.4\ in^2$.

$$s = \frac{A_v f_y d_v}{V_s}$$

$$= \frac{(0.4\ in^2)\left(60\frac{kips}{in^2}\right)(37.3\ in)}{136.0\ kips}$$

$$= 6.58\ in$$

Use s = 6.5 in

Minimum transverse reinforcement (where b_v is the width of the web) shall be

A Eq. 5.8.2.5-1

$$A_{v,\,min} \geq 0.0316\sqrt{f'_c}\,\frac{b_v s}{f_y}$$

$$A_{v,\,min} = 0.0316\sqrt{4.5\frac{kips}{in^2}}\,\frac{(18\ in)(6.5\ in)}{\left(60\frac{kips}{in^2}\right)}$$

$$= 0.13\ in^2 < 0.4\ in^2 \text{ at 6.5 in spacing provided [OK]}$$

At the critical section (x = 3.58 ft) from the bearing of the left support, the transverse reinforcement will be two no. 4 bar stirrups at 6.5 in spacing.

Determine the maximum spacing for transverse reinforcement required.

A Art. 5.8.2.7

v_u is the average factored shear stress in the concrete.

A Eq. 5.8.2.9-1

$$v_u = \frac{V_u}{\Phi b_v d_v}$$

$$= \frac{203.4\ kips}{(0.9)(18\ in)(37.3\ in)}$$

$$= 0.337\frac{kips}{in^2}$$

If $v_u < 0.125\ f'_c$

$$s_{max} = 0.8\ d_v < 24\ \text{in}$$

A Eq. 5.8.2.7-1

$$v_u = 0.337\frac{\text{kips}}{\text{in}^2} < 0.125\left(4.5\frac{\text{kips}}{\text{in}^2}\right) = 0.563\frac{\text{kips}}{\text{in}^2}$$

Thus, $s_{max} = 0.8(37.3\ \text{in}) = 29.8\ \text{in} \geq 24\ \text{in}$
$s_{provided} = 6.5\ \text{in} \leq 24\ \text{in}$ [OK]

Check tensile capacity of longitudinal reinforcement.

A Art. 5.8.3.5

Shear causes tension in the longitudinal reinforcement in addition to shear caused by the moment.

A Art. 5.8.3.5

At the critical section for shear, x is 3.58 ft.

M_u	factored moment	711.59 ft-kips
V_u	factored shear	203.4 kips
Φ_f, Φ_v	resistance factors for moment and shear	0.90

A Art. 5.5.4.2.1

For two no. 4 stirrups at 6.5 in spacing and cot 45° = 1.0, the shear resistance provided by shear reinforcement is,

A Eq. 5.8.3.3-4

$$V_s = \frac{A_v f_y d_v}{s} = \frac{\left(0.4\ \text{in}^2\right)\left(60\frac{\text{kips}}{\text{in}^2}\right)\left(37.3\ \text{in}\right)}{6.5\ \text{in}}$$

$$= 137.7\ \text{kips}$$

The required tensile capacity of the reinforcement on the flexural tensile side shall satisfy:

A Eq. 5.8.3.5-1

$$A_s f_y = \frac{M_u}{\varphi_f d_v} + \left(\frac{V_u}{\varphi_v} - 0.5V_s\right)\cot\theta$$

$$= \frac{711.59\ \text{ft-kips}}{(0.9)(37.3\ \text{in})}\left(12\frac{\text{in}}{\text{ft}}\right) + \left(\frac{203.4\ \text{kips}}{0.9} - 0.5(137.7\ \text{kips})\right)(1.0)$$

$$= 411.5\ \text{kips}$$

The available tension capacity of longitudinal reinforcement is

$$A_s f_y = \left(18.75\ \text{in}^2\right)\left(60\frac{\text{kips}}{\text{in}^2}\right)$$

$$= 1123.2\ \text{kips} > 411.5\ \text{kips [OK]}$$

Step 2: Check the Fatigue Limit State

A Art. 5.5.3

The fatigue strength of the bridge is related to the range of live load stress and the number of stress cycles under service load conditions. The fatigue load combination specified in AASHTO Table 3.4.1-1 is used to determine the allowable fatigue stress range, f_f (= constant-amplitude fatigue threshold, $(\Delta F)_{TH}$) (AASHTO Art. 5.5.3.1). The minimum live-load stress level, f_{min}, is determined by combining the fatigue load with the permanent loads. f_{min} will be positive if it is in tension.

The factored load, Q, is calculated using the load factors given in AASHTO Tables 3.4.1-1 and 3.4.1-2.

For a simple span bridge with no prestressing, there are no compressive stresses in the bottom of the beam under typical dead load conditions. Therefore, fatigue must be considered.

A Art. 5.5.3.1

Fatigue Limit State II (related to finite load-induced fatigue life) Q = 0.75(LL + IM), where LL is the vehicular live load and IM is the dynamic load allowance.

A Tbl. 3.4.1-1

The following information will be used to determine the fatigue load.

One design truck that has a constant spacing of 30 ft between 32 kip axles.

A Art. 3.6.1.4.1

Dynamic load allowance, IM, is 15%.

A Tbl. 3.6.2.1-1

A distribution factor, DFM, for one traffic lane shall be used.

A Art. 3.6.1.4.3b

The multiple presence factor of 1.2 shall be removed.

A Comm. 3.6.1.1.2

The allowable fatigue stress range, f_f (= $(\Delta F)_{TH}$), shall be equal to (24 – 0.33 f_{min}).

A Eq. 5.5.3.2-1

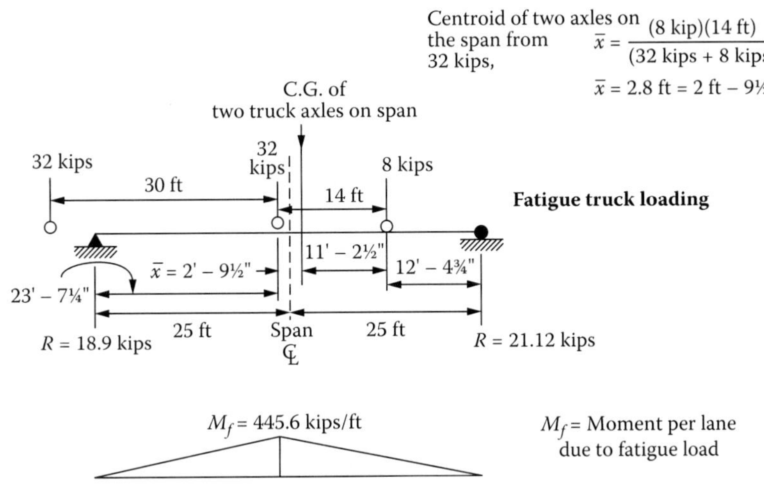

FIGURE 2.14
Fatigue truck loading and maximum moment at 32 kips position per lane due to fatigue loading.

f_f Allowable fatigue stress range, kips/in^2

f_{min} Minimum live-load stress resulting from the fatigue load combined with the permanent loads; positive if tension, kips/in^2

See Figure 2.14.

The moment per lane due to fatigue load, M_f, is 445.6 ft-kips. For one traffic lane loaded,

$$DFM_{si} = 0.629 \text{ lanes for interior girders}$$

$$DFM_{se} = 0.87 \text{ lanes for exterior girders}$$

The multiple presence factor of 1.2 must be removed. Therefore, the distribution factor for fatigue load using DFM_{se} is,

A Art. 3.6.1.1.2

$$DFM_{fatigue} = \frac{0.87}{1.2} = 0.725 \text{ lanes}$$

Find the unfactored fatigue load moment per beam.

The moment due to fatigue load per beam, $M_{fatigue}$, is

$$M_{fatigue} = M_f(DFM)(1 + IM)$$

$$= (445.6 \text{ kip-ft})(0.725)(1 + 0.15)$$

$$= 371.5 \text{ kip-ft}$$

Fatigue Investigations.

It is noted that provisions used for Fatigue I are conservative for Fatigue II load.

A Art. 5.5.3.1

For fatigue investigations, the section properties shall be based on cracked sections when the sum of the stresses, due to unfactored permanent loads and Fatigue I load combination, is tensile and exceeds $0.095\sqrt{f_c'}$. Referring to the unfactored exterior girder moments previously found,

$$M_{DC} + M_{DW} + M_{fatigue} = 756.3 \text{ kip-ft} + 84.4 \text{ kip-ft} + 371.5 \text{ kip-ft}$$

$$= 1212.2 \text{ kip-ft [for exterior girders]}$$

Find the tensile stress at the bottom fiber of gross section of interior girders using the stresses for exterior girders to be conservative.

A Art. 5.5.3.1

$$S_g = \frac{I_g}{y_t}$$

$$= \frac{264{,}183.6 \text{ in}^4}{31.4 \text{ in}}$$

$$= 8{,}413.5 \text{ in}^3$$

The tensile stress at the bottom fiber f_t is

$$f_t = \frac{M}{S_g}$$

$$= \frac{1212.2 \text{ kip-ft}\left(\frac{12 \text{ in}}{1 \text{ ft}}\right)}{8413.5 \text{ in}^3}$$

$$= 1.73 \text{ kips/in}^2$$

$$0.095\sqrt{f_c'} = 0.095\sqrt{4.5\frac{\text{kips}}{\text{in}^2}}$$

$$= 0.20 \text{ kips/in}^2 < f_t = 1.73 \text{ kip/in}^2$$

A Art. 5.5.3.1

Therefore, cracked section analysis should be used for fatigue investigation. The modulus of elasticity for concrete with $w_c = 0.15$ kips/in² is

$$E_c = 33{,}000(w_c)^{1.5}\sqrt{f'_c}$$

A Eq. 5.4.2.4-1

$$= 33{,}000\left(0.15\frac{\text{kips}}{\text{ft}^3}\right)^{1.5}\sqrt{4.5\frac{\text{kips}}{\text{in}^2}}$$

$$E_c = 4{,}066.8 \text{ kips/in}^2$$

The modulus of elasticity for steel, E_s, is 29,000 kips/in²
The modular ratio between steel and concrete is

$$n = \frac{E_s}{E_c}$$

$$= 7.13, \text{ use } n = 7$$

Determine the transformed area.

$$nA_s = 7(18.75 \text{ in}^2) = 131.25 \text{ in}^2$$

Find the factored fatigue moment per beam.

The factored load for Fatigue II load combination, Q, is

A Table 3.4.1-1

$$Q = 0.75(LL + IM)$$

$$M_{F,\,fatigue} = 0.75(M_{fatigue}) = 0.75(371.5 \text{ kip-ft}) = 278.6 \text{ kip-ft}$$

See Figure 2.15 for the cracked section analysis.

Find the distance, x, from the top fiber to the neutral axis. Taking the moment of areas about the neutral axis,

$$(120 \text{ in})(x)\left(\frac{x}{2}\right) = (131.25 \text{ in}^2)(38.5 \text{ in} - x)$$

$$x = 8.15 \text{ in}$$

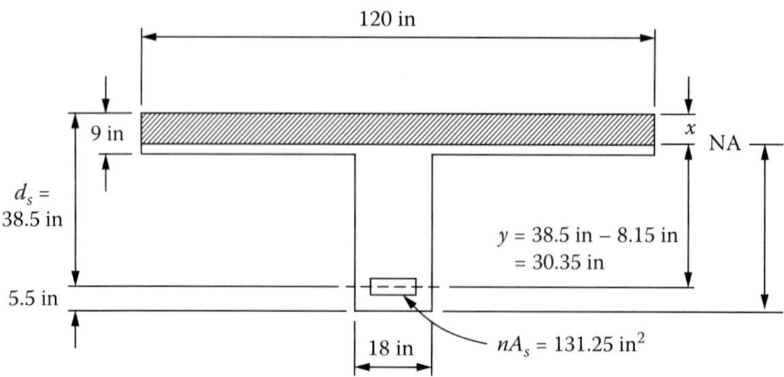

FIGURE 2.15
Cracked section determination of T-beam.

$$I_{NA} = \frac{1}{3}bx^3 + Ay^2$$

$$= \frac{1}{3}(120\text{ in})(8.15\text{ in})^3 + (131.25\text{ in}^2)(38.5\text{ in} - 8.15\text{ in})^2$$

$$= 142{,}551.0\text{ in}^4$$

Find the stress in the reinforcement.

The stress in the reinforcement due to the factored fatigue live load f_s, is,

$$f_s = n\left(\frac{(M)(y)}{I}\right)$$

$$= n\left(\frac{M_{F\text{ fatigue}}}{I_{NA}}\right)(y)$$

$$= (7)\left(\frac{(278.6\text{ kip-ft})\left(\frac{12\text{ in}}{1\text{ ft}}\right)}{142{,}551\text{ in}^4}\right)(38.5\text{ in} - 8.15\text{ in})$$

$$f_s = 4.98\text{ kips/in}^2$$

Find the fatigue stress range.

The permissible stress range in the reinforcing bars resulting from the fatigue load combination, f_f, must not exceed the stress in the reinforcement, f_s.

Using the previously calculated unfactored exterior girder moments due to the dead loads, the total moment due to the dead load is

$$M_{dead\ load} = M_{DC} + M_{DW} = 756.3 \text{ ft-kips} + 84.4 \text{ ft-kips}$$

$$= 840.7 \text{ ft-kips}$$

The minimum live load stress resulting from the fatigue load is

$$f_{min} = f_{s,\ dead\ load} = (n)\left(\frac{M_{deadload}y}{I}\right)$$

$$= (7)\left(\frac{(840.7 \text{ ft-kips})\left(12\frac{\text{ft}}{\text{in}}\right)(38.5 \text{ in} - 8.15 \text{ in})}{142551 \text{ in}^4}\right)$$

$$= 15.0 \text{ kips/in}^2$$

The allowable fatigue stress range f_f, is,

A Eq. 5.5.3.2-1

$$f_f = 24 - 0.33\, f_{min}$$

$$= 24 - (0.33)(15.0 \text{ kips/in}^2)$$

$$= 19.05 \text{ kips/in}^2$$

f_s = stress in the reinforcement due to the factored fatigue live load

$= 4.98 \text{ kips/in}^2 < f_f = 19.05 \text{ kips/in}^2$ [OK]

The stress in the reinforcement due to the fatigue live load is less than the allowable fatigue stress range, so it is ok.

Step 3: Check the Service I Limit State

The factored load, Q, is calculated using the load factors given in AASHTO Table 3.4.1-1.

The load combination relating to the normal operational use of the bridge with all loads taken at their nominal values (with no impact allowance, no load factors, and so on), is similar to that for the allowable stress methods. This combination is for deflection control and crack width control in reinforced concrete.

$$Q = 1.0(DC) + 1.0(DW) + 1.0(TL + LN)$$

A Tbl. 3.4.1-1

Although deflection and depth limitations are optional in AASHTO, bridges should be designed to avoid undesirable structural or psychological effects due to deformations.

A Art. 2.5.2.6.1

Criteria for deflection in this section shall be considered optional. However, in the absence of other criteria, the following limits may be considered.

A Art. 2.5.2.6.2

For steel, aluminum, concrete vehicular bridges,
  vehicular load general — span/800
For wood vehicular bridges,
  vehicular and pedestrian loads — span/425
  vehicular load on wood planks and panels
  (extreme relative deflection between adjacent edges) — 0.10 in.

Using the previously calculated exterior girder moments due to the dead and live loads, the total service load moment is

$$M_{service} = 1.0\ M_{DC} + 1.0\ M_{DW} + 1.0(M_{TL} + M_{LN})$$

$$= (1.0)(756.3\ \text{ft-kips}) + (1.0)(84.4\ \text{ft-kips})$$

$$+ (1.0)(717.4\ \text{ft-kips} + 174.0\ \text{ft-kips})$$

$$= 1732.1\ \text{ft-kips}$$

Find the control of cracking by distribution of reinforcement.

The maximum spacing of tension reinforcement applies if the tension in the cross section exceeds 80% of the modulus of rupture, f_r, at the Service I Limit State.

A Art. 5.7.3.4

The modulus of rupture is

A Art. 5.4.2.6

$$f_r = 0.24\sqrt{f'_c}$$

$$= 0.24\sqrt{4.5\frac{\text{kips}}{\text{in}^2}}$$

$$= 0.51 \text{ kips/in}^2$$

The section modulus at the bottom fiber for the gross cross section (where y_t is equivalent to y_b) is

$$S_g = \frac{I_g}{y_t}$$

where:

y_t = distance from the neutral axis to the extreme tension fiber
= y + 5.5 in = 35.85 in

$$= \frac{264{,}183.6 \text{ in}^4}{35.85 \text{ in}}$$

$$= 7369.1 \text{ in}^3$$

The tensile stress at the bottom fiber is

$$f_t = \frac{M_{service}}{S_g} = \frac{(1732.1 \text{ kip-ft})\left(\dfrac{12 \text{ in}}{1 \text{ ft}}\right)}{7369.1 \text{ in}^3}$$

$$= 2.82 \text{ kips/in}^2 > 0.8\, f_r = 0.8(0.51 \text{ kips/in}^2) = 0.41 \text{ kips/in}^2$$

Thus, flexural cracking is controlled by limiting the spacing, s, in the tension reinforcement. Therefore, the maximum spacing, s, of the tension reinforcement should satisfy the following.

A Eq. 5.7.3.4-1

$$s \le \frac{700\,\gamma_e}{\beta_s f_{ss}} - 2\,d_c$$

The exposure factor for a class 2 exposure condition for concrete, γ_e, is 0.75. The overall thickness, h, is 44 in. The concrete cover measured from extreme tension fiber to the center of the flexural reinforcement located closest thereto is

A Art. 5.7.3.4

$$d_c = 1.5 \text{ in} + 0.5 \text{ in} + 1.41 \text{ in}/2$$

$$= 2.7 \text{ in (see Figure 2.11)}$$

The modular ratio between the steel and concrete, n, was previously calculated as 7. The tensile stress in steel reinforcement at the Service I Limit State is

A Art. 5.7.3.4

$$f_{ss} = n\left(\frac{M_{service}}{S_g}\right),$$

where:

$$S_g = \frac{I_g}{y} = \frac{264{,}183.6 \text{ in}^4}{30.35 \text{ in}} = 8704.6 \text{ in}^3$$

$$= 7\left(\frac{(1732.1 \text{ kip-ft})\left(\frac{12 \text{ in}}{1 \text{ ft}}\right)}{8{,}413.5 \text{ in}^3}\right)$$

$$= 16.7 \text{ kips/in}^2$$

The ratio of the flexural strain at extreme tension face to the strain at the centroid of reinforcement layer nearest to the tension face is

A Art. 5.7.3.4

$$\beta_s = 1 + \left(\frac{d_c}{0.7(h - d_c)}\right)$$

$$= 1 + \left(\frac{2.7 \text{ in}}{0.7(44 \text{ in} - 2.7 \text{ in})}\right)$$

$$= 1.09$$

Therefore, the maximum spacing, s, of the tension reinforcement shall satisfy the following.

$$s \leq \frac{700\,\gamma_e}{\beta_s f_{ss}} - 2\,d_c$$

$$\leq \frac{700(0.75)}{(1.09)\left(16.7\frac{\text{kips}}{\text{in}^2}\right)} - 2(2.7 \text{ in})$$

$$\leq 24.3 \text{ in}$$

Provided bar spacing, s, is 4.19 in, which is less than the maximum spacing allowed so it is ok (see Figure 2.11).

Step 4: Design the Deck Slab

In AASHTO, concrete decks can be designed either by the empirical method or the traditional method.

A Art. 9.7

The empirical design method may be used for concrete deck slabs supported by longitudinal components if the following conditions are satisfied.

A Arts. 9.7.2, 9.7.2.4

a. The supporting components are made of steel and/or concrete.
b. The deck is fully cast-in-place and water cured.
c. The deck is of uniform depth, except for haunches at girder flanges and other local thickening.
d. The ratio of effective length to design depth does not exceed 18 and is not less than 6.0.

The effective length is face-to-face of the beams monolithic with slab; therefore,

$$\text{ratio} = \frac{\text{effective length}}{\text{design depth}} = \frac{10\ \text{ft} - 1.5\ \text{ft}}{(9\ \text{in})\left(\dfrac{1\ \text{ft}}{12\ \text{in}}\right)}$$

$$= 11.33 < 18 > 6\ [\text{OK}]$$

e. Core depth of the slab (out-to-out of reinforcement) is not less than 4 in.

Assuming clear cover of 2.5 in for top bars and 1.5 in for bottom bars,

$$9\ \text{in} - 2.5\ \text{in} - 1.5\ \text{in} = 5\ \text{in} > 4\ \text{in}\ [\text{OK}]$$

f. The effective length does not exceed 13.5 ft.

$$8.5\ \text{ft} < 13.5\ \text{ft}\ [\text{OK}]$$

g. The minimum depth of slab is not less than 7 in.

$$9 \text{ in} > 7 \text{ in [OK]}$$

h. There is an overhang beyond the centerline of the outside girder of at least 5 times the depth of slab. This condition is satisfied if the overhang is at least 3 times the depth of slab and a structurally continuous concrete barrier is made composite with the overhang.

Overhang = (4.0 ft)(12 in/ft).

Check requirements.

$$(5)(9 \text{ in}) = 45 \text{ in} < 48 \text{ in [OK]}$$

$$(3)(9 \text{ in}) = 27 \text{ in} < 48 \text{ in [OK]}$$

i. The specified 28-day strength of the deck concrete, f'_c, is not less than 4 kips/in².

$$f'_c = 4.5 \text{ kips/in}^2 > 4 \text{ kips/in}^2 \text{ [OK]}.$$

j. The deck is made composite with the supporting structural elements.

Extending the beam stem stirrups into deck will satisfy this requirement [OK].

The empirical method may be used because all of the above conditions are met.

Four layers of isotropic reinforcement shall be provided. The minimum amount of reinforcement in each direction shall be 0.27 in²/ft for bottom steel, and 0.18 in²/ft for top steel. Spacing of steel shall not exceed 18 in.

A Art. 9.7.2.5

Bottom reinforcement: no. 5 bars at 12 in; $A_s = 0.31 \text{ in}^2 > 0.18 \text{ in}^2$ [OK].

Top reinforcement: no. 5 bars at 18 in; $A_s = 0.20 \text{ in}^2 > 0.18 \text{ in}^2$ [OK].

The outermost layers shall be placed in the direction of the slab length.

Alternatively, the traditional design method may be used.

A Art. 9.7.3

If the conditions for the empirical design are not met, or the designer chooses not to use the empirical method, the LRFD method allows the use of the traditional design method.

Concrete slab shall have four layers of reinforcement, two in each direction and clear cover shall comply with the following AASHTO Art. 9.7.1.1 provisions.

Top bars have 2.5 in and bottom bars have 1.5 in clear covers.

The approximate method of analysis for decks divides the deck into strips perpendicular to the support elements. The width of the strip is calculated according to AASHTO Art. 4.6.2.1.3.

A Art. 4.6.2.1

The width of the primary strip is calculated for a cast-in-place concrete deck using the following calculations from AASHTO Table 4.6.2.1.3-1.

Overhang: 45 + 10 X

+M: 26 + 6.6 S

–M: 48 + 3.0 S

The spacing of supporting elements, S, is measured in feet. The distance from load to point of support, X, is measured in feet.

A deck slab may be considered as a one-way slab system. The strip model of the slab consists of the continuous beam, and the design truck is positioned transverse for the most critical actions with a pair of 16 kip axles spaced 6 ft apart.

Design Example 2: Load Rating of Reinforced Concrete T-Beam by the Load and Resistance Factor Rating (LRFR) Method

Problem Statement

Use the *Manual for Bridge Evaluation* (*MBE-2*), Second Edition 2011, Section 6: Load Rating, Part A Load and Resistance Factor Rating (LRFR). Determine the load rating of the reinforced-concrete T-beam bridge interior beam using the bridge data given for the Limit State Combination Strength I. Also refer to *FHWA July 2009 Bridge Inspection System*, Condition and Appraisal rating guidelines.

L	span length	50 ft
s	beam spacing	10 ft
f'_c	concrete strength	4.5 kips/in^2
f_y	specified minimum yield strength of steel	60 kips/in^2
DW	future wearing surface load	0.03 kips/ft^2
ADTT	average daily truck traffic in one direction	1900
	skew	0°
Year Built:		1960

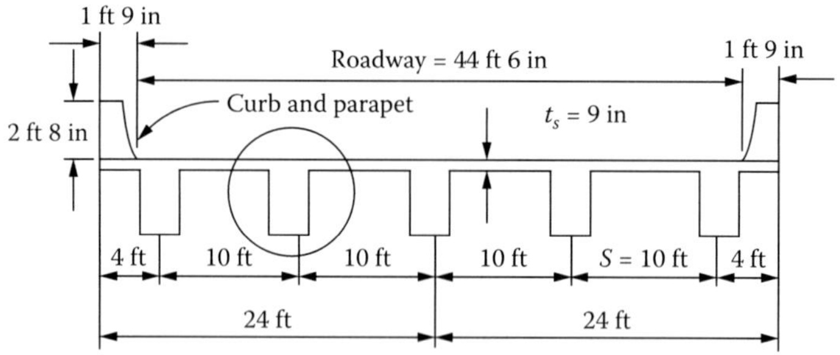

FIGURE 2.16
T-beam bridge cross section.

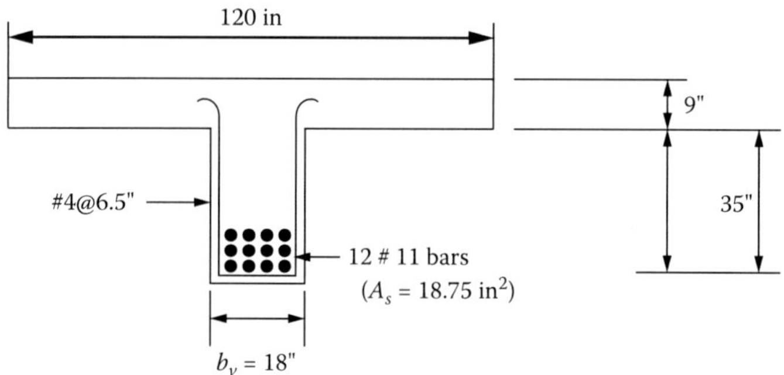

FIGURE 2.17
T-beam section.

Condition: Good condition with some minor problems
NBIS Item 59, Condition Good (Code 7)

See Figures 2.16 and 2.17.

Solution

Step 1: Analysis of Dead Load Components—Interior Beam

1.A Components and Attachments – DC

$$DC_{\text{T-beam – Structural}} = 1.78 \text{ kips/ft}$$

$$DC_{\text{Curbs \& Parapet}} = 0.202 \text{ kips/ft}$$

$$\text{Total DC per beam} = DC_{\text{T-beam – Structural}} + DC_{\text{Curbs \& Parapet}}$$

$$= 1.78 \text{ kips/ft} + 0.202 \text{ kips/ft}$$

$$\text{Total DC per beam} = 1.98 \text{ kips/ft}$$

$$M_{DC} = \frac{w_{DC}L^2}{8} = \frac{\left(1.98\,\frac{\text{kips}}{\text{ft}}\right)(50\text{ ft})^2}{8} = 618.75\text{ kip-ft}$$

$$V_{DC} = \frac{w_{DC}L}{2} = \frac{\left(1.98\,\frac{\text{kips}}{\text{ft}}\right)(50\text{ ft})}{2} = 49.5\text{ kips}$$

1.B Wearing Surface – DW

$$w_{DW} = \left(0.03\,\frac{\text{kips}}{\text{ft}^2}\right)\left(\frac{10\text{ ft}}{\text{beam}}\right) = 0.3\,\frac{\text{kips}}{\text{ft}}$$

$$M_{DW} = \frac{w_{DW}L^2}{8} = \frac{\left(0.3\,\frac{\text{kips}}{\text{ft}}\right)(50\text{ ft})^2}{8} = 93.75\text{ kip-ft}$$

$$V_{DW} = \frac{w_{DW}L}{2} = \frac{\left(0.3\,\frac{\text{kips}}{\text{ft}}\right)(50\text{ ft})}{2} = 7.5\text{ kips}$$

Step 2: Analysis of Live Load Components—Interior Beam

2.A Live Load Distribution Factor

AASHTO Cross-Section Type (e)

A Tbl 4.6.2.2.1-1

For preliminary design,

A Tbl. 4.6.2.2.1-2

$$\left[\frac{K_g}{12\,Lt_s^3}\right]^{0.1} = 1.05$$

2.A.1 Distribution Factor for Moment in Interior Beam, DFM

A Tbl 4.6.2.2b-1 or Appendix A

i) One lane loaded, DFM_{si}

$$DFM_{si} = 0.06 + \left[\frac{10\text{ ft}}{14}\right]^{0.4}\left[\frac{10\text{ ft}}{50\text{ ft}}\right]^{0.3}[1.05]$$

$$DFM_{si} = 0.626\text{ lanes}$$

ii) Two or more lanes loaded, DFM_{mi}

$$DFM_{mi} = 0.075 + \left[\frac{S}{9.5}\right]^{0.6}\left[\frac{S}{L}\right]^{0.2}\left[\frac{K_g}{12\,Lt_s^3}\right]^{0.1}$$

$$DFM_{mi} = 0.075 + \left[\frac{10\text{ ft}}{9.5}\right]^{0.6}\left[\frac{10\text{ ft}}{50\text{ ft}}\right]^{0.2}[1.05]$$

$$DFM_{mi} = 0.860\text{ lanes [controls]}$$

2.A.2 Distribution Factor for Shear in Interior Beam, DFV

AASHTO [Tbl 4.6.2.2.3a-1] or Appendix C

i) One lane loaded, DFV_{si}

$$DFV_{si} = 0.36 + \frac{S}{25}$$

$$DFV_{si} = 0.36 + \frac{10\text{ ft}}{25}$$

$$DFV_{si} = 0.76\text{ lanes}$$

ii) Two or more lanes loaded, DFV_{mi}

$$DFV_{mi} = 0.2 + \frac{S}{12} - \left(\frac{S}{35}\right)^{2.0}$$

$$DFV_{mi} = 0.2 + \frac{10\text{ ft}}{12} - \left(\frac{10\text{ ft}}{35}\right)^{2.0}$$

$$DFV_{mi} = 0.95\text{ lanes [controls]}$$

2.B Maximum Live Load Effects

2.B.1 Maximum Live Load (HL-93) Moment at Midspan (see Design Example 1)

Design Truck Moment = M_{tr} = 620.0 kip-ft

Design Lane Moment = M_{lane} = 200.0 kip-ft

Design Tandem Axles Moment = 575.0 kip-ft

IM = 33%

A Tbl. 3.6.2.1-1

$$M_{LL+IM} = M_{tr}(1 + IM) + M_{lane}$$

$$= 620.0 \text{ kip-ft}(1 + 0.33) + 200.0 \text{ kip-ft}$$

$$M_{LL+IM} = 1024.6 \text{ kip-ft}$$

2.B.2 Distributed live load moment per beam

$$M_{LL+IM\ per\ beam} = (M_{LL+IM})(DFM)$$

$$= (1024.6 \text{ kip-ft})(0.860)$$

$$M_{LL+IM\ per\ beam} = 881.16 \text{ kip-ft}$$

Step 3: Determine the Nominal Flexural Resistance, M_n

3.A Effective Flange Width of T-Beam, b_e

Effective flange width is the width of the deck over girders.

A Comm. 4.6.2.6.1

Average spacing of beams = (10 ft)(12 in/1 ft) = 120.0 in

Use b_e = 120.0 in

3.B Determination of Nominal Flexural Resistance in Bending, M_n, is

A Eq. 5.7.3.2.2-1

$$M_n = A_s f_y \left(d_s - \frac{a}{2} \right)$$

A_s for 12 - #11 bars = 18.75 in^2

$$b_e = 120 \text{ in}$$

$$\beta_1 = 0.85 - \left(\frac{f'_c - 4000}{1000}\right)(0.05)$$

A Art. 5.7.2.2

$$\beta_1 = 0.85 - \left(\frac{4500\frac{\text{lbf}}{\text{in}^2} - 4000}{1000}\right)(0.05)$$

$$\beta_1 = 0.825$$

$$c = \frac{A_s f_y}{0.85\, f'_c\, \beta_1 b}$$

A Eq. 5.7.3.1.1-4

$$c = \frac{(18.75 \text{ in}^2)\left(60\frac{\text{kips}}{\text{in}^2}\right)}{0.85\left(4.5\frac{\text{kips}}{\text{in}^2}\right)(0.825)(120 \text{ in})}$$

$$c = 2.97 \text{ in} < t_s = 9 \text{ in}$$

$$a = c\beta$$

A Art. 5.7.2.2

$$a = (2.97 \text{ in})(0.825)$$

$$a = 2.45 \text{ in}$$

d_s distance from extreme compression fiber to the centroid of the tensile reinforcement 38.5 in

See Figure 2.18.

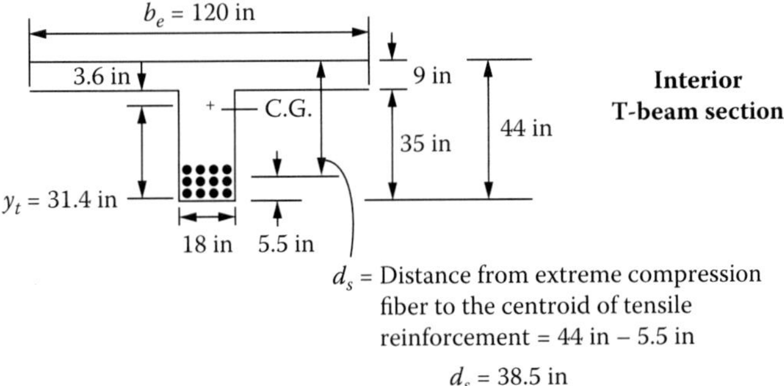

FIGURE 2.18
Interior T-beam section for determination of flexural resistance.

$$M_n = A_s f_y \left[d_s - \frac{a}{2} \right]$$

$$M_n = \left(18.75\ \text{in}^2\right)\left(60\frac{\text{kips}}{\text{in}^2}\right)\left[38.5\ \text{in} - \frac{2.45\ \text{in}}{2}\right]$$

$$M_n = 41934\ \text{kip-in}(1\ \text{ft}/12\ \text{in})$$

$$M_n = 3494.5\ \text{kip-ft}$$

Step 4: Determine the Shear at the Critical Section near Support

A Art 5.8.3.2

Vertical Stirrups: #4 bars at 6.5 in spacing

$$A_v = 2A_{\#4\ \text{bar}} = 2\left(\frac{\pi}{4}\text{x}\left(\frac{4}{8}\text{in}\right)^2\right) = 0.40\ \text{in}^2$$

The location of the critical section for shear shall be taken as d_v from the bearing face of the support.

The effective shear depth, d_v, can be determined as

A Eq. C.5.8.2.9-1; 5.8.2.9

$$d_v = \frac{M_n}{A_s f_y}$$

$$d_v = \frac{(3494.5 \text{ kip-ft})\left(\dfrac{12 \text{ in}}{1 \text{ ft}}\right)}{(18.74 \text{ in}^2)\left(60 \dfrac{\text{kips}}{\text{in}^2}\right)}$$

$$d_v = 37.3 \text{ in [controls]}$$

But d_v need not be taken less than to be less than the greater of 0.9 d_e or 0.72 h:

A Art. 5.8.2.9

$$d_v = 0.9\ d_e (= d_s)$$

$$= 0.9(38.5 \text{ in})$$

$$d_v = 34.7 \text{ in}$$

or

$$d_v = 0.72\ h$$

$$= 0.72(44 \text{ in})$$

$$d_v = 31.7 \text{ in}$$

Use d_v = 37.3 in.

Check: d_v = distance between the resultants of the tensile and compressive force due to flexure,

$$d_v = d_s - \frac{a}{2} = 38.5 \text{ in} - \frac{2.45 \text{ in}}{2}$$

$$= 37.3 \text{ in (check OK)}$$

See Figure 2.19.
The critical section for shear from face of bearing support, d_v, is 37.3 in.

Calculate shear at x = 37.3 in +5.7 in = 43 in from the center of bearing.

Shear at x = (43 in) (1 ft/12 in) = 3.58 ft from the support end.

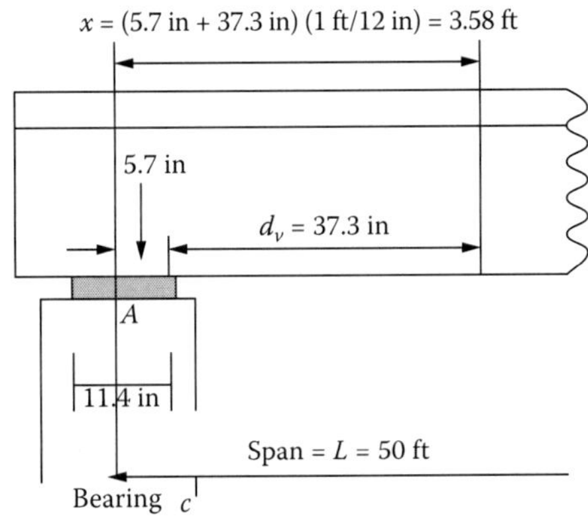

FIGURE 2.19
Critical section for shear at support.

Dead Load

$$w_{DC} = 1.98 \text{ kips/ft}$$

$$w_{DW} = 0.3 \text{ kips/ft}$$

Maximum live load shear at critical section on a simple span

MBE-2 App. Tbl. H6B

Live load shear at critical section due to design truck load, HS-20,

Let x = L – x; multiply by 2 for two wheel lines.

$$\text{HS-20: } V = \frac{36 \text{ kips}(x - 9.33)}{L} x2 \text{ and } x = (L\text{-}x)$$

$$V = \frac{72 \text{ kips}\left[(L - x) - 9.33\right]}{L}$$

$V_{tr = HS\text{-}20}$ at x = 3.58 ft:

$$V_{tr=HS\text{-}20} = \frac{72 \text{ kips}\left[(50 \text{ ft} - 3.58 \text{ ft}) - 9.33\right]}{50 \text{ ft}}$$

$$V_{tr = HS\text{-}20} = 53.41 \text{ kips per lane}$$

$$V_{TL} = (V_{tr})(DFV)(1 + IM)$$

$$= (53.41 \text{ kips per lane})(0.95)(1.33)$$

$$V_{TL} = 67.5 \text{ kips per beam}$$

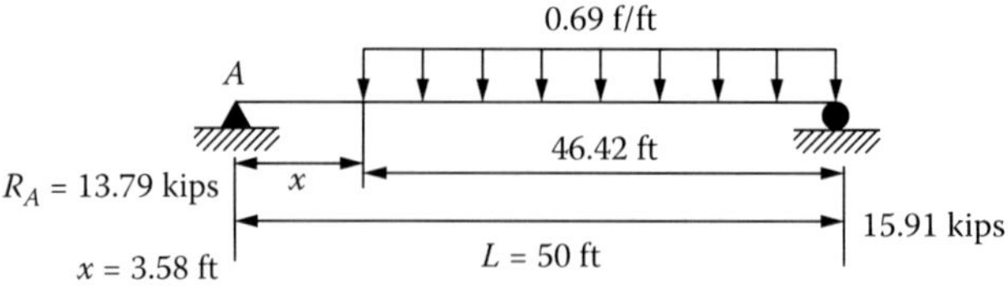

FIGURE 2.20
Live load shear at critical shear section due to lane load.

Live load shear at critical shear section due to lane load (see Figure 2.20),

$$R_A = \frac{\left(0.64\,\frac{\text{kips}}{\text{ft}}\right)(46.42\text{ ft})\left(\frac{46.42\text{ ft}}{2}\right)}{50\text{ ft}}$$

$$R_A = 13.79\text{ kips}$$

$$V_{\text{ln@x = 3.58 ft}} = 13.79\text{ kips}$$

$$V_{LN} = V_{ln}(DFV) = (13.79\text{ kips})(0.95) = 13.1\text{ kips per beam}$$

$$V_{LL+IM} = \text{total live load shear} = 67.5\text{ kips} + 13.1\text{ kips} = 80.6\text{ kips}$$

Nominal Shear Resistance, V_n, shall be the lesser of:

A Art. 5.8.3.3

$$V_n = V_c + V_s$$

A Eq. 5.8.3.3-1, A Eq. 5.8.3.3-2

$$V_n = 0.25\,f_c'\,b_v d_v$$

$$V_c = 0.0316\,\beta\sqrt{f_c'}\,b_v d_v$$

$$V_s = \frac{A_v f_y d_v \cot\theta}{s}$$

$$\alpha = 90^\circ$$

A Comm. 5.8.3.3-1

For simplified procedure,

A Art. 5.8.3.4.1

$$\beta = 2.0;\ \theta = 45°$$

$$V_c = 0.0316(2.0)\sqrt{4.5\frac{kips}{in^2}}(18\ in)(37.3\ in)$$

$$V_c = 90.0\ kips$$

$$V_s = \frac{(0.40\ in^2)\left(60\frac{kips}{in^2}\right)(37.3\ in)\cot 45°}{6.5\ in}$$

$$V_s = 137.7\ kips$$

$$V_n = 90.0\ kips + 137.7\ kips = 227.7\ kips\ [controls]$$

$$V_n = 0.25\ f_c'\ b_v d_v$$

$$V_n = 0.25\left(4.5\frac{kips}{in^2}\right)(18\ in)(37.3\ in)$$

$$V_n = 356.1\ kips$$

Step 5: Summary of Capacities and Demands for Interior Concrete T-Beam

See Table 2.3.

Step 6: Calculation of T-Beam Load Rating

MBE-2 Eq. 6A.4.2.1-1

$$RF = \frac{C-(\gamma_{DC})(DC)-(\gamma_{DW})(DW)\pm(\gamma_P)(P)}{(\gamma_{LL})(LL+IM)}$$

TABLE 2.3

Dead Loads and Distributed Live Loads Effects Summary for Interior T-Beam

	DC	DW	LL Distribution Factor	Distributed LL+Impact	Nominal Capacity
Moment, kip-ft	618.75	93.75	0.860	881.16	3494.5
Shear, kips[a]	42.4	6.4	0.95	80.6	227.7

[a] At critical section (= 3.58 ft from bearing).

For the strength limit states:

MBE-2 Eq. 6A.4.2.1-2

$$C = \Phi_c \Phi_s \Phi R_n$$

where:
RF = rating factor
C = capacity
f_R = allowable stress specified in the LRFD code
R_n = nominal member resistance (as inspected)
DC = dead load effect due to structural components and attachments
DW = dead load effect due to wearing surface and utilities
P = permanent loads other than dead loads
LL = live load effect
IM = dynamic load allowance
γ_{DC} = LRFD load factor for structural components and attachments
γ_{DW} = LRFD load factor for wearing surfaces and utilities
γ_p = LRFD load factor for permanent loads other than dead loads = 1.0
γ_{LL} = evaluation live load factor
Φ_c = condition factor

MBE-2 Tbl 6A.4.2.3-1

Φ_c = 1.0 for good condition
Φ_s = system factor

MBE-2 Tbl 6A.4.2.4-1

Φ_s = 1.0 for all other girder bridges
Φ = LRFD resistance factor

A Art. 5.5.4.2.1

Φ = 0.90 for shear and flexure

6.A Strength I Limit State

$$RF = \frac{(\Phi_c)(\Phi_s)(\Phi)R_n - (\gamma_{DC})(DC) - (\gamma_{DW})(DW)}{(\gamma_{LL})(LL + IM)}$$

MBE-2 Eq. 6A.4.2-1

Please see Table 2.4.

TABLE 2.4 (MBE-2 Tbl 6A.4.2.2-1)

Load Factors for Load Rating for Reinforced Concrete Bridge

				Design Live Load	
Bridge Type	**Limit State**	**Dead Load, γ_{DC}**	**Dead Load, γ_{DW}**	**Inventory γ_{LL}**	**Operating γ_{LL}**
Reinforced Concrete	Strength I	1.25	1.5	1.75	1.35

6.B Load Rating for Inventory Level

Flexure:

$$RF = \frac{\left[(1.0)(1.0)(0.9)(3494.5\text{ kip-ft}) - (1.25)(618.75\text{ kip-ft}) - (1.50)(93.75\text{ kip-ft})\right]}{(1.75)(881.16\text{ kip-ft})}$$

$$RF = \frac{2231.0\text{ kip-ft}}{1542.03\text{ kip-ft}}$$

$$RF = 1.45$$

Shear:

$$RF = \frac{(1.0)(1.0)(0.9)(227.0\text{ kips}) - (1.25)(42.4\text{ kips}) - (1.50)(6.4\text{ kips})}{(1.75)(80.6\text{ kips})}$$

$$RF = \frac{141.7\text{ kips}}{141.05\text{ kips}}$$

$$RF = 1.00$$

6.C Operating Level Load Rating

Flexure:

$$RF = (1.45)\left(\frac{1.75}{1.35}\right) = 1.88$$

Shear:

$$RF = (1.00)\left(\frac{1.75}{1.35}\right) = 1.30$$

6.D Summary of Rating Factors for Load and Resistance Factor Rating Method (LRFR)—Interior Beam for Limit State Strength I

Please see Table 2.5.

TABLE 2.5

Rating Factor (RF)

		Design Load Rating	
Limit State	**Force Effect**	**Inventory**	**Operating**
Strength I	Flexure	1.45	1.88
	Shear	1.00	1.30

6.E Loading in Tons, RT

MBE-2 Eq. 6A.4.4-1

$$RT = RF \times W$$

where:
RF = rating factor (Table 2.5)
RT = rating in tons for trucks used in computing live load effect
W = weight in tons of truck used in computing live load effect

---

Design Example 3: Composite Steel–Concrete Bridge

Situation

BW	barrier weight	0.5 kips/ft
f'_c	concrete strength	4 kips/in^2
ADTT	average daily truck traffic in one direction	2500
	design fatigue life	75 yr
w_{FWS}	future wearing surface load	25 lbf/ft^2
d_e	distance from the centerline of the exterior web of exterior beam to the interior edge of curb or traffic barrier	2 ft
L	span length	40 ft
	load of stay-in-place metal forms	7 lbf/ft^2
S	beam spacing	8 ft
w_c	concrete unit weight	150 lbf/ft^3
f_y	specified minimum yield strength of steel	60 kips/in^2
t_s	slab thickness	8 in

Requirements

Design the superstructure of a composite steel–concrete bridge for the load combinations Strength I, Service II, and Fatigue II Limit States using the given design specifications.

Solution

AISC 13th Ed. Tbl. 1-1*

For the bare steel section, W24x76, please see Figure 2.21.

* Refers to the American Institute of Steel Construction.

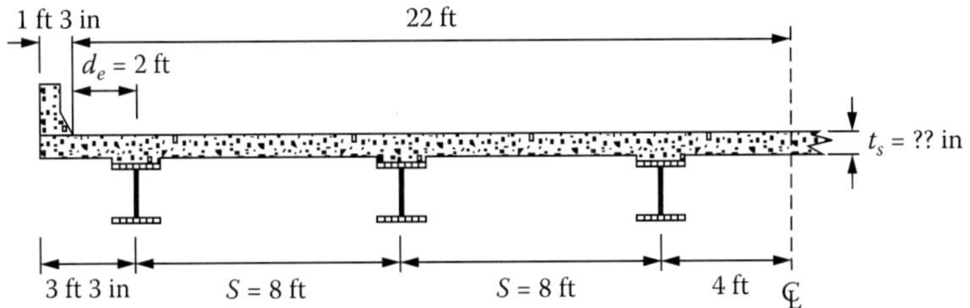

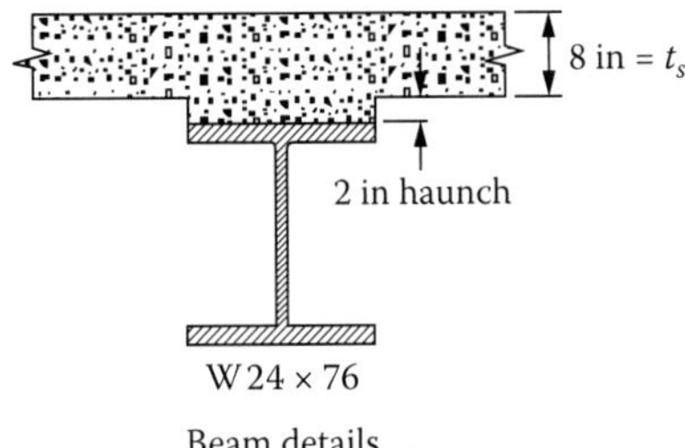

FIGURE 2.21
Composite steel–concrete bridge example.

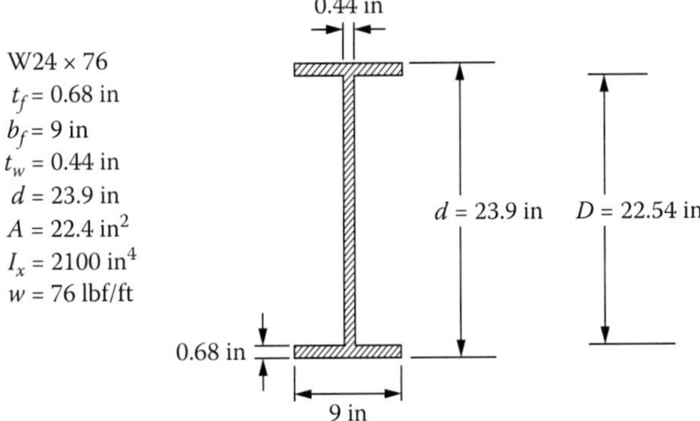

FIGURE 2.22
Steel section.

Find the effective flange widths

A Art. 4.6.2.6

The effective flange width for interior beams is equal to the beam spacing (see Figure 2.22)

$$b_{e,int} = (8\text{ ft})(12\text{ in/ft}) = 96\text{ in}$$

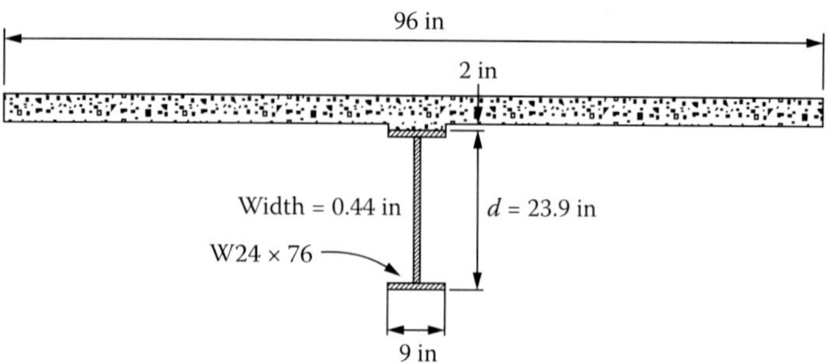

FIGURE 2.23
Composite steel section.

The effective flange width for exterior beams is equal to half of the adjacent beam spacing plus the overhang width

$$b_{e,ext} = S/2 + \text{overhang} = (96\text{ in})/2 + 39\text{ in} = 87\text{ in}$$

Find the dead load moments and shears. See Figure 2.23.
Find the noncomposite dead load, DC_1, per interior girder.

$$DC_{slab} = (96\text{ in})\left(\frac{1\text{ ft}}{12\text{ in}}\right)(8\text{ in})\left(\frac{1\text{ ft}}{12\text{ in}}\right)\left(0.15\frac{\text{kips}}{\text{ft}^3}\right) = 0.8\text{ kips/ft}$$

$$DC_{haunch} = (2\text{ in})\left(\frac{1\text{ ft}}{12\text{ in}}\right)(9\text{ in})\left(\frac{1\text{ ft}}{12\text{ in}}\right)\left(0.15\frac{\text{kips}}{\text{ft}^3}\right) = 0.02\text{ kips/ft}$$

Assuming 5% of steel weight for diaphragms, stiffeners, and so on,

$$DC_{steel} = 0.076\frac{\text{kips}}{\text{ft}} + (0.05)\left(0.076\frac{\text{kips}}{\text{ft}}\right) = 0.08\text{kips/ft}$$

$$DC_{\text{stay-in-place forms}} = \left(7\frac{\text{lbf}}{\text{ft}^2}\right)(\text{roadway width})\left(\frac{1}{6\text{ girders}}\right)$$

$$= \left(0.007\frac{\text{kips}}{\text{ft}^2}\right)(44\text{ ft})\left(\frac{1}{6}\right)$$

$$= 0.051\text{ kips/ft}$$

The noncomposite dead load per interior girder is

$$DC_1 = DC_{slab} + DC_{haunch} + DC_{steel} + DC_{stay\text{-}in\text{-}place\ forms}$$

$$= 0.8\ \text{kips/ft} + 0.02\ \text{kips/ft} + 0.08\ \text{kips/ft} + 0.051\ \text{kips/ft}$$

$$= 0.951\ \text{kips/ft}$$

The shear for the noncomposite dead load is

$$V_{DC_1} = \frac{wL}{2} = \frac{\left(0.951\frac{\text{kips}}{\text{ft}}\right)(40\ \text{ft})}{2} = 19.02\ \text{kips}$$

The moment for the noncomposite dead load is

$$M_{DC_1} = \frac{wL^2}{8} = \frac{\left(0.951\frac{\text{kips}}{\text{ft}}\right)(40\ \text{ft})^2}{8} = 190.2\ \text{ft-kips}$$

Find the composite dead load, DC_2, per interior girder.

$$DC_2 = DC_{barrier} = \frac{\left(0.5\frac{\text{kips}}{\text{ft}}\right)(2\ \text{barriers})}{6\ \text{girders}} = 0.167\ \text{kips/ft}$$

The shear for the composite dead load per interior girder is

$$V_{DC_2} = \frac{wL}{2} = \frac{\left(0.167\frac{\text{kips}}{\text{ft}}\right)(40\ \text{ft})}{2} = 3.34\ \text{kips}$$

The moment for the composite dead load per interior girder is

$$M_{DC_2} = \frac{wL^2}{8} = \frac{\left(0.167\frac{\text{kips}}{\text{ft}}\right)(40\ \text{ft})^2}{8} = 33.4\ \text{ft-kips}$$

The shear for the total composite dead load for interior girders is

$$V_{DC} = V_{DC1} + V_{DC2} = 19.02\ \text{ft-kips} + 3.34\ \text{ft-kips} = 22.36\ \text{kips}$$

The moment for the total composite dead load for interior girders is

$$M_{DC} = M_{DC1} + M_{DC2} = 190.2 \text{ ft-kips} + 33.4 \text{ ft-kips} = 223.6 \text{ ft-kips}$$

The future wearing surface dead load, DW, for interior girders is

$$DW_{FWS} = \frac{w_{FWS}L}{\text{no. of beams}} = \frac{\left(25\frac{\text{lbf}}{\text{ft}^2}\right)(44 \text{ ft})}{6 \text{ beams}} = 0.183 \text{ kips/ft}$$

The shear for the wearing surface dead load for interior girders is

$$V_{DW} = \frac{wL}{2} = \frac{\left(0.183\frac{\text{kips}}{\text{ft}}\right)(40 \text{ ft})}{2} = 3.66 \text{ kips}$$

The moment for the wearing surface dead load for interior girders is

$$M_{DW} = \frac{wL^2}{8} = \frac{\left(0.183\frac{\text{kips}}{\text{ft}}\right)(40 \text{ ft})^2}{8} = 36.6 \text{ ft-kips}$$

Assume that the dead load for the exterior girders is the same as for interior girders. This is conservative.

Dead load summary

$V_{DC} = 22.4$ kips
$M_{DC} = 223.6$ ft-kips
$V_{DW} = 3.66$ kips
$M_{DW} = 36.6$ ft-kips

Find the live loads.
Find the unfactored moments, unfactored shears, and distribution factors for the live loads.

A Art. 4.6.2.2.1

The modulus of elasticity of concrete where $w_c = 150 \text{ lbf/ft}^3$ (0.15 kips/ft³) is

A Eq. 5.4.2.4-1

$$E_c = (33{,}000)w_c^{1.5}\sqrt{f'_c} = (33{,}000)\left(0.15\frac{\text{kips}}{\text{ft}^3}\right)^{1.5}\sqrt{4\frac{\text{kips}}{\text{in}^2}}$$

$$= 3834 \text{ kips/in}^2$$

The modulus of elasticity for the steel, E_s, is 29,000 kips/in^2.

The modular ratio between steel and concrete is

$$n = \frac{E_s}{E_c} = \frac{29{,}000\dfrac{\text{kips}}{\text{in}^2}}{3834\dfrac{\text{kips}}{\text{in}^2}} = 7.56$$

Use n = 8.

The distance between the centers of gravity of the deck and beam, e_g, is 17.95 in.

AISC 13th Ed. Tbl. 1-1

The moment of inertia for the beam, I, is 2100 in^4.

The area, A, is 22.4 in^2.

Using the previously calculated values and the illustration shown, calculate the longitudinal stiffness parameter, K_g. See Figure 2.24.

The longitudinal stiffness parameter, K_g, is

A Eq. 4.6.2.2.1-1

$$K_g = n(I + Ae_g^2) = (8)(2100\ \text{in}^4 + (22.4\ \text{in}^2)(17.95\ \text{in})^2)$$

$$= 74{,}539\ \text{in}^4$$

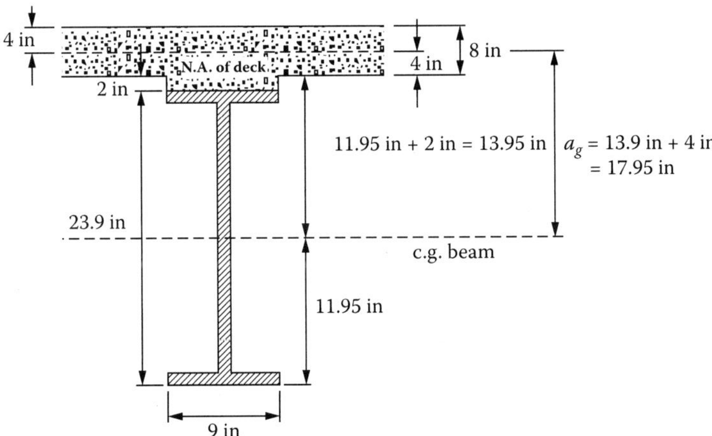

FIGURE 2.24
Composite section for stiffness parameter, K_g.

For cross-section type (a):

A Tbl. 4.6.2.2.1-1

$$S = 8 \text{ ft}$$

$$L = 40 \text{ ft}$$

$$t_s = 8 \text{ in}$$

The distribution factor for moments for interior girders with one design lane loaded is

A Tbl. 4.6.2.2.2b-1

$$DFM_{si} = 0.06 + \left(\frac{S}{14}\right)^{0.4}\left(\frac{S}{L}\right)^{0.3}\left(\frac{K_g}{12\,Lt_s^3}\right)^{0.1}$$

$$= 0.06 + \left(\frac{8 \text{ ft}}{14}\right)^{0.4}\left(\frac{8 \text{ ft}}{40 \text{ ft}}\right)^{0.3}\left(\frac{74{,}539 \text{ in}^4}{12(40 \text{ ft})(8 \text{ in})^3}\right)^{0.1}$$

$$= 0.498 \text{ lanes}$$

The distribution factor for moments for interior girders with two or more design lanes loaded is

$$DFM_{mi} = 0.075 + \left(\frac{S}{9.5}\right)^{0.6}\left(\frac{S}{L}\right)^{0.2}\left(\frac{K_g}{12\,Lt_s^3}\right)^{0.1}$$

$$= 0.075 + \left(\frac{8 \text{ ft}}{9.5}\right)^{0.6}\left(\frac{8 \text{ ft}}{40 \text{ ft}}\right)^{0.2}\left(\frac{74{,}539 \text{ in}^4}{12(40 \text{ ft})(8 \text{ in})^3}\right)^{0.1}$$

$$= 0.655 \text{ lanes [controls]}$$

The distribution factor for shear for interior girders with one design lane loaded is

A Tbl. 4.6.2.2.3a-1

$$DFV_{si} = 0.36 + \left(\frac{S}{25}\right) = 0.36 + \left(\frac{8 \text{ ft}}{25}\right)$$

$$= 0.68 \text{ lanes}$$

The distribution factor for shear for interior girders with two or more design lanes loaded is

$$\mathrm{DFV_{mi}} = 0.2 + \frac{S}{12} - \left(\frac{S}{35}\right)^2 = 0.2 + \frac{8\ \mathrm{ft}}{12} - \left(\frac{8\ \mathrm{ft}}{35}\right)^2$$

$$= 0.814 \text{ lanes [controls]}$$

The distribution factor for moments for exterior girders with one design lane loaded can be found using the lever rule.

A Art. 6.1.3.1; A Tbl. 4.6.2.2.2d-1 or Appendix B

The multiple presence factors, m, must be applied where the lever rule is used. Please see Figure 2.25.

A Comm. 3.6.1.1.2; Tbl. 3.6.1.1.2-1

$$\Sigma \mathrm{M_{@hinge}} = 0$$

$$= \left(\frac{P}{2}\right)(8\ \mathrm{ft}) + \left(\frac{P}{2}\right)(2\ \mathrm{ft}) - R(8\ \mathrm{ft})$$

$$R = 0.625\ P = 0.625$$

A multiple presence factor, m, for one lane loaded is applied to the distribution factor for moments of exterior girders.

A Tbl. 3.6.1.1.2-1

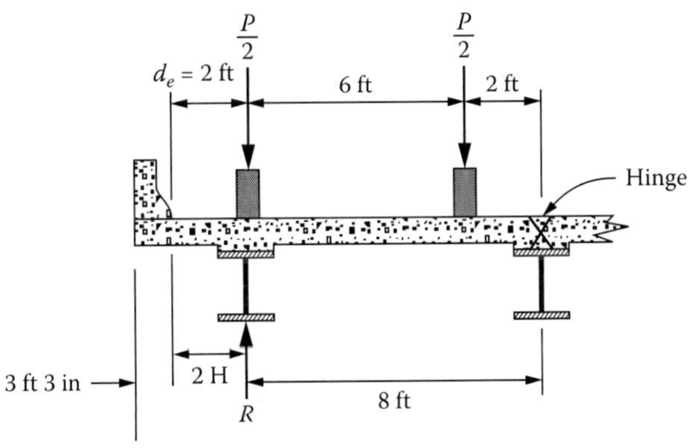

FIGURE 2.25
Lever rule for determination of distribution factor for moment in exterior beam, one lane loaded.

$$DFM_{se} = (DFM_{se})\,(m) = R(m)$$

$$DFM_{se} = (0.625)(1.2) = 0.75 \text{ lanes [controls]}$$

The distribution factor for moments for exterior girders with two or more design lanes loaded using g, a distribution factor is

A Tbl. 4.6.2.2.2d-1 or Appendix B

The distribution factor for interior girder, g_{int}, is 0.655.

The correction factor for distribution, where the distance from the centerline of the exterior web of the exterior beam to the curb, d_e, is 2 ft, is

$$e = 0.77 + \frac{d_e}{9.1} = 0.77 + \frac{2\text{ ft}}{9.1} = 0.9898$$

$$DFM_{me} = g = (e)(g_{int}) = (0.9898)(0.655) = 0.648 \text{ lanes [controls]}$$

Or, stated another way,

$$DFM_{me} = (e)DFM_{mi} = (0.9898)(0.655) = 0.648 \text{ lanes}$$

The distribution factor for moments for exterior girders with one design lane loaded is 0.75 (> 0.648).

Use the lever rule to find the distribution factor for shear in exterior girders for one design lane loaded. This is the same as DFM_{se} for one design lane loaded (= 0.75).

A Tbl. 4.6.2.2.3b-1 or Appendix D

$$DFV_{se} = 0.75$$

Find the distribution factor for shear for exterior girders with two or more design lanes loaded using g, a distribution factor.

$$e = 0.6 + \frac{d_e}{10} = 0.6 + \frac{2\text{ft}}{10} = 0.8$$

$$g = (e)(g_{int}) = (0.8)(0.814) = 0.6512 \text{ lanes}$$

Or, stated another way,

$$DFV_{me} = e(DFV_{mi}) = (0.8)(0.814) = 0.6512 \text{ lanes}$$

Find the unfactored live load effects with dynamic load allowance, IM.

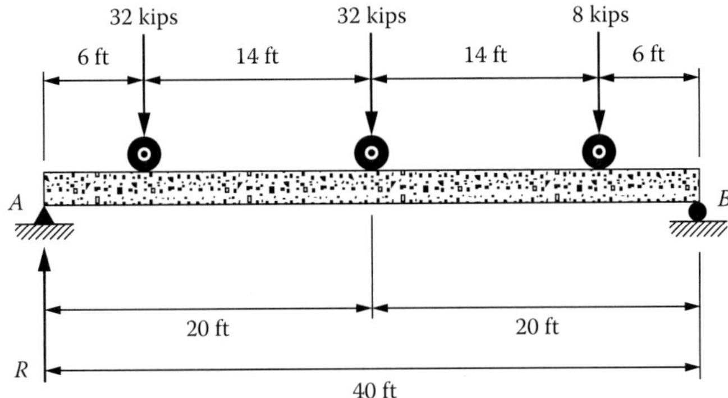

FIGURE 2.26
Load position for moment at midspan for design truck load (HS-20).

The HL-93 live load is made of either the design truck or the design tandem load (whichever is larger) and the design lane load.

A Art. 3.6.1.2

Find the design truck moment due to the live load. See Figure 2.26.

$$\Sigma M_{@B} = 0$$

$$R_A(40\text{ ft}) = (32\text{ kips})(34\text{ ft}) + (32\text{ kips})(20\text{ ft}) + (8\text{ kips})(6\text{ ft})$$

$$R_A = 44.4\text{ kips}$$

The design truck (HS-20) moment due to live load is

$$M_{tr} = (44.4\text{ kips})(20\text{ ft}) - (32\text{ kips})(14\text{ ft}) = 440\text{ ft-kips}$$

Find the design tandem moment due to the live load. Please see Figure 2.27.

$$\Sigma M_{@B} = 0$$

$$R_A = \left(\frac{25\text{ kips}}{2}\right) + (25\text{ kips})\left(\frac{16\text{ ft}}{40\text{ ft}}\right)$$

$$R_A = 22.5\text{ kips}$$

The design tandem moment due to the live load is

$$M_{tandem} = (22.5\text{ kips})(20\text{ ft}) = 450\text{ ft-kips [controls]}$$

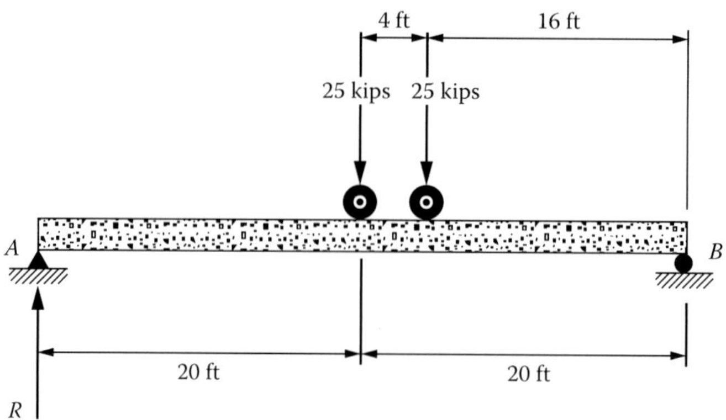

FIGURE 2.27
Load position for moment at midspan for design tandem load.

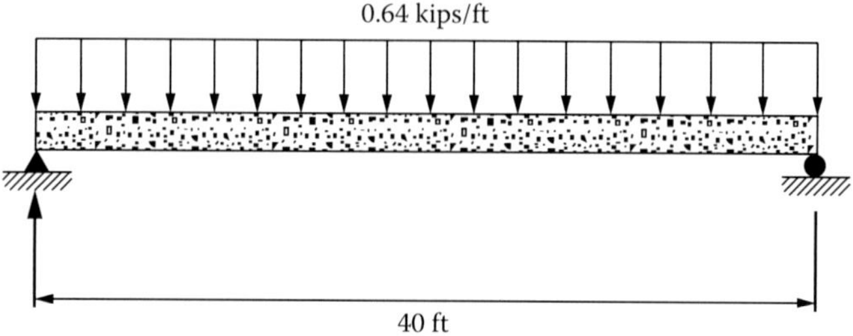

FIGURE 2.28
Load position for moment at midspan for design lane load.

Find the design lane moment due to the live load. Please see Figure 2.28.

$$M_{ln} = \frac{wL^2}{8} = \frac{\left(0.64 \frac{\text{kips}}{\text{ft}}\right)(40\ \text{ft})^2}{8} = 128\ \text{ft-kips}$$

The dynamic load allowance, IM, is 33% (applied to the design truck or the design tandem only, not to the design lane load).

A Art. 3.6.2.1

The total unfactored live load moment is

$$M_{LL+IM} = M_{tandem}(1 + IM) + M_{ln} = (450\ \text{ft-kips})(1 + 0.33) + 128\ \text{ft-kips}$$

$$= 726.5\ \text{ft-kips per lane}$$

Find the shear due to the design truck live load. Please see Figure 2.29.

$$\Sigma M_{@B} = 0$$

$$R_A(40\text{ ft}) = (32\text{ kips})(40\text{ ft}) + (32\text{ kips})(26\text{ ft}) + (8\text{ kips})(12\text{ ft})$$

$$R_A = 55.2\text{ kips (controls)}$$

The shear due to design truck load, V_{tr}, is 55.2 kips [controls].

Find the shear due to the design tandem load. Please see Figure 2.30.

$$\Sigma M_{@B} = 0$$

$$R_A(40\text{ ft}) = (25\text{ kips})(40\text{ ft}) + (25\text{ kips})(36\text{ ft})$$

$$R_A = 47.5\text{ kips}$$

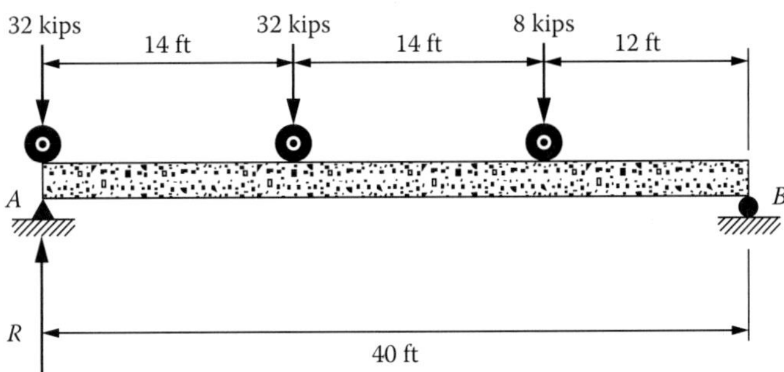

FIGURE 2.29
Load position for shear at support for design truck load (HS-20).

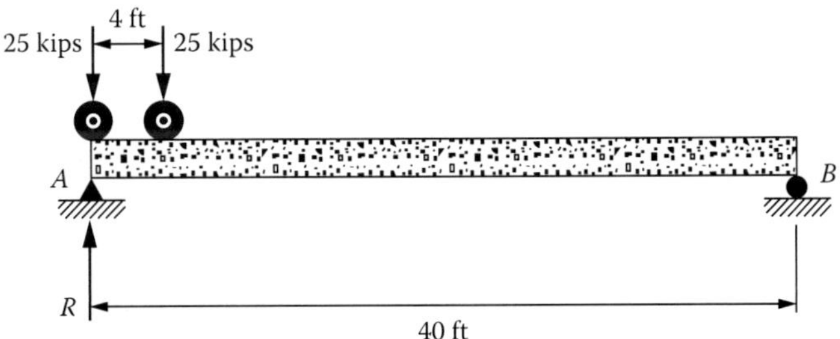

FIGURE 2.30
Load position for shear at support design tandem load.

The shear due to the design tandem load, V_{tandem}, is 47.5 kips.

Find the shear due to the design lane load. Please see Figure 2.31.

$$V = V_{ln} = \frac{wL}{2} = \frac{\left(0.64\frac{\text{kips}}{\text{ft}}\right)(40\text{ ft})}{2} = 12.8\text{ kips}$$

The dynamic load allowance, IM, is 33% (applied to the design truck or the design tandem only, but not to the design lane load).

The total unfactored live load shear is

$$V_{LL+IM} = V_{tr}(1 + IM) + V_{ln} = (55.2\text{ kips})(1.33) + 12.8\text{ kips}$$

$$= 86.22\text{ kips per lane}$$

Please see Table 2.6.

Find the factored moments and shears.

A Tbls. 3.4.1-1; 3.4.1-2

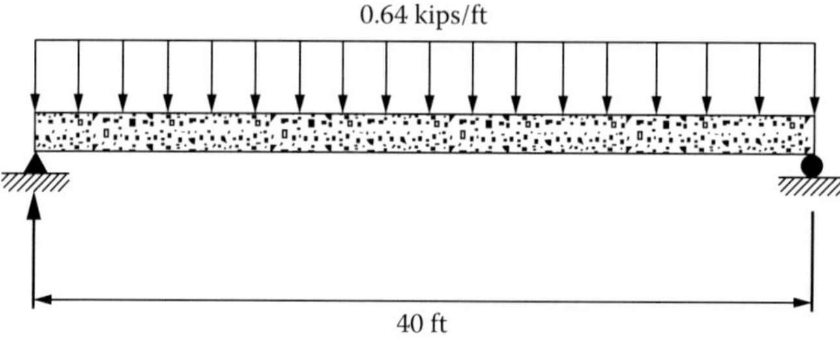

FIGURE 2.31
Load position for shear at support for design lane load.

TABLE 2.6

Complete Live Load Effect Summary

Girder Location	No. of Design Lanes Loaded	Unfactored M_{LL+IM} (ft-kips per Lane)	DFM	Unfactored M_{LL+IM} (ft-kips per Girder)	Unfactored V_{LL+IM} (kips per Lane)	DFV	Unfactored V_{LL+IM} (kips per Girder)
Interior	1	726.5	0.498	—	86.22	0.68	—
	2	726.5	0.655	475.9	86.22	0.814	70.2
Exterior	1	726.5	0.75	544.9	86.22	0.75	64.7
	2	726.5	0.648	—	86.22	0.6512	—

Check the Strength I Limit State.

$$U = 1.25\ DC + 1.5\ DW + 1.75(LL + IM)$$

The factored moment for interior girders is

$$M_u = 1.25\ M_{DC} + 1.5\ M_{DW} + 1.75\ M_{LL+IM}$$

$$= (1.25)(223.6\text{ ft-kips}) + (1.5)(36.6\text{ ft-kips}) + (1.75)(475.9\text{ ft-kips})$$

$$= 1167.2\text{ ft-kips}$$

The factored shear for interior girders is

$$V_u = 1.25\ V_{DC} + 1.5\ V_{DW} + 1.75\ V_{LL+IM}$$

$$= (1.25)(22.4\text{ kips}) + (1.5)(3.66\text{ kips}) + (1.75)(70.2\text{ kips})$$

$$= 156.3\text{ kips}$$

The factored moment for exterior girders is

$$M_u = (1.25)(223.6\text{ ft-kips}) + (1.5)(36.6\text{ ft-kips}) + (1.75)(544.9\text{ ft-kips})$$

$$= 1288\text{ ft-kips}$$

The factored shear for interior girders is

$$V_u = (1.25)(22.4\text{ kips}) + (1.5)(3.66\text{ kips}) + (1.75)(64.7\text{ kips})$$

$$= 146.7\text{ kips}$$

Find the shear and moment capacity of the composite steel–concrete section. Find the plastic moment capacity, M_p, for interior girders. See Figure 2.32.

A App. D6.1; Tbl. D6.1-1

$b_{s,ext}$	effective flange width for exterior beam	87 in
$b_{s,int}$	effective flange width for interior beam	96 in
b_c	flange width, compression	9 in
D	clear distance between flanges	22.54 in
F_{yc}	specified minimum yield strength of the compression flange	60 kips/in^2
F_{yt}	specified minimum yield strength of the tension flange	60 kips/in^2

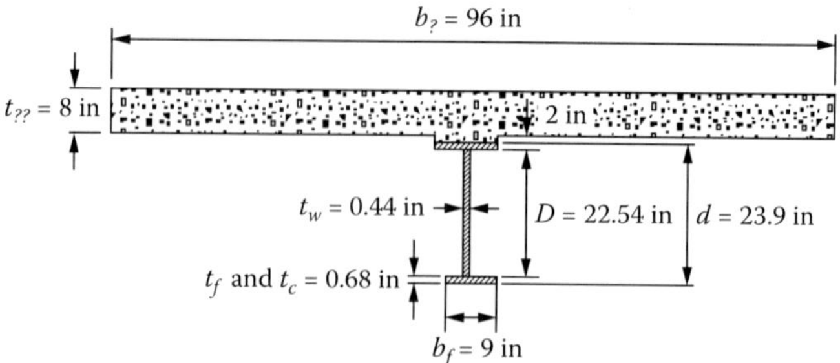

Assume longitudinal reinforcement as shown.

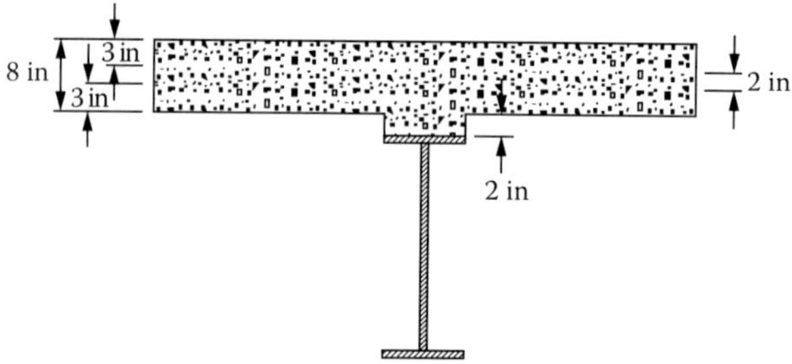

FIGURE 2.32
Composite steel–concrete section for shear and moment capacity calculation.

F_{yw}	specified minimum yield strength of the web	60 kips/in^2
$t_c = t_f$	flange thickness, compression	0.68 in
t_s	thickness of the slab	8 in
$t_t = t_f$	flange thickness, tension flange	0.68 in
t_w	web thickness	0.44 in
C_{rt}	distance from top of concrete deck to top layer of longitudinal concrete deck reinforcement	3 in
C_{rb}	distance from top of concrete deck to bottom layer of longitudinal concrete deck reinforcement	5 in
P_{rb}	plastic force in the bottom layer of longitudinal deck reinforcement	
P_{rt}	plastic force in the top layer of longitudinal deck reinforcement	

Locate the plastic neutral axis (PNA).

A App. D6-1; Tbl. D6.1-1; D6.1-2

The plastic compressive force in the slab is

$$P_s = 0.85\,f'_c b_s t_s$$

$$= (0.85)(4.5\text{ kips/in}^2)(96\text{ in})(8\text{ in})$$

$$= 2937.6\text{ kips}$$

Neglect strength of slab reinforcement.

$$P_{rb} = P_{rt} = 0\ \text{[conservative]}$$

The plastic force in the compression flange is

$$P_c = F_{yc} b_c t_c = (60\text{ ksi})(9\text{ in})(0.68\text{ in}) = 367.2\text{ kips}$$

The plastic force in the web is

$$P_w = F_{yw} D t_w = (60\text{ ksi})(22.54\text{ in})(0.44\text{ in}) = 595\text{ kips}$$

The plastic force in the tension flange is

$$P_t = F_{yt} b_t t_t = (60\text{ ksi})(9\text{ in})(0.68\text{ in}) = 367.2\text{ kips}$$

Case I (in web): $P_t + P_w \geq P_c + P_s$

$$P_t + P_w = 367.2\text{ kips} + 595\text{ kips} = 962.2\text{ kips}$$

$$P_c + P_s = 367.2\text{ kips} + 2937.6\text{ kips} = 3305\text{ kips}$$

$$962.2\text{ kips} < 3305\text{ kips [no good]}$$

Case II (in top flange): $P_t + P_w + P_c \geq P_s$

A App. Tbl. D6.1-1

$$P_t + P_w + P_c = 367.2\text{ kips} + 595\text{ kips} + 367.2\text{ kips} = 1329.4\text{ kips}$$

$$P_s = 2937.6\text{ kips}$$

$$1329.4\text{ kips} < 2937.6\text{ kips [no good]}$$

Case III (in deck, below bottom reinforcement): $P_t + P_w + P_c \geq (C_{rb}/t_s)P_s$

$$P_t + P_w + P_c = 367.2 \text{ kips} + 595 \text{ kips} + 367.2 \text{ kips} = 1329.4 \text{ kips}$$

$$(C_{rb}/t_s)P_s = (5 \text{ in}/8 \text{ in})(2937.6) = 1836 \text{ kips}$$

$$1329.4 \text{ kips} < 1836 \text{ kips [no good]}$$

Case IV (not considered because the effects of slab reinforcement are ignored: $P_{rb} = 0$ and $P_{rt} = 0$)

Case V (in deck, between reinforcement layers): $P_t + P_w + P_c \geq (C_{rb}/t_s)P_s$

$$P_t + P_w + P_c = 367.2 \text{ kips} + 595 \text{ kips} + 367.2 \text{ kips} = 1329.4 \text{ kips}$$

$$(C_{rt}/t_s)P_s = (3 \text{ in}/8 \text{ in})(2937.6) = 1102 \text{ kips}$$

$$1329.4 \text{ kips} > 1102 \text{ kips [OK]}$$

The plastic neutral axis (PNA) is between the top and bottom layers of reinforcement, The PNA, $\bar{Y}$, for Case V is,

A App. Tbl. D6.1-1

$$\bar{Y} = t_s\left(\frac{P_{rb} + P_c + P_w + P_t - P_{rt}}{P_s}\right)$$

$$= (8 \text{ in})\left(\frac{0 \text{ kips} + 367.2 \text{ kips} + 595 \text{ kips} + 367.2 \text{ kips} - 0 \text{ kips}}{2937.6 \text{ kips}}\right)$$

$$= 3.62 \text{ in from top of deck}$$

The distance from the mid-depth of steel compression flange to the PNA is

$$d_c = 8 \text{ in} + 2 \text{ in} + (0.68 \text{ in})/2 - 3.62 \text{ in} = 6.72 \text{ in}$$

The distance from the mid-depth of steel web to the PNA is

$$d_w = 8 \text{ in} + 2 \text{ in} + (23.9 \text{ in})/2 - 3.62 \text{ in} = 18.33 \text{ in}$$

The distance from the mid-depth of steel tension flange to the PNA is

$$d_t = 8 \text{ in} + 2 \text{ in} + 23.9 \text{ in} - (0.68 \text{ in})/2 - 3.62 \text{ in} = 29.94 \text{ in}$$

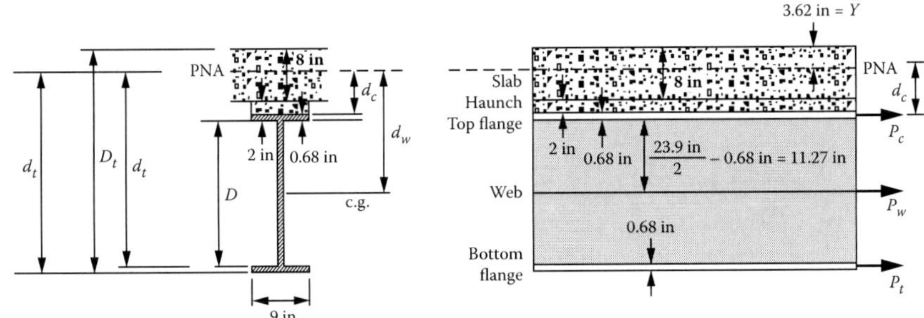

Section and cross section of interior girder showing plastic neutral axis location

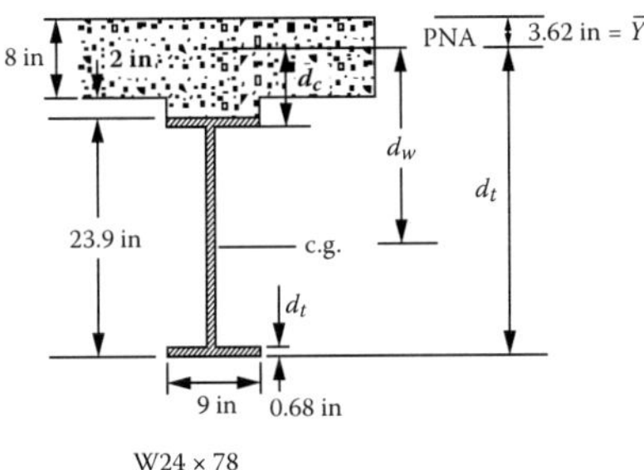

FIGURE 2.33
Section and cross section of interior girder for plastic moment capacity.

See Figure 2.33.

The plastic moment capacity, M_p, for Case V in interior girders is

A Tbl. D6.1-1

$$M_p = \left(\frac{\bar{Y}^2 P_s}{2t_s}\right) + \left(P_{rt}d_{rt} + P_{rb}d_{rb} + P_c d_c + P_w d_w + P_t d_t\right)$$

$$= \left(\frac{(3.62 \text{ in})^2 (2937.6 \text{ kips})}{2(8 \text{ in})}\right) + \begin{pmatrix} 0 + 0 + (367.2 \text{ kips})(6.72 \text{ in}) \\ +(631 \text{ kips})(18.33 \text{ in}) \\ +(367.2 \text{ kips})(29.94 \text{ in}) \end{pmatrix}$$

$$= 2406 \text{ in-kips} + 2467.6 \text{ in-kips} + 11566.2 \text{ in-kips} + 10994 \text{ in-kips}$$

$$= 27{,}432 \text{ in-kips}(1 \text{ ft}/12 \text{ in})$$

$$= 2286 \text{ ft-kips}$$

NOTE: Neglect strength of slab reinforcement ($P_{rt} = P_{rb} = 0$).

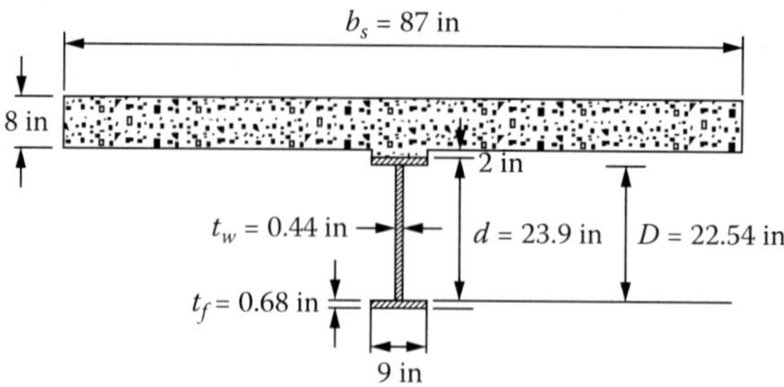

FIGURE 2.34
Composite cross section for exterior beam.

Find the plastic moment capacity, M_p, for Case V in exterior girders. Please see Figure 2.34.

$b_{s,ext}$	effective flange width for exterior beam	87 in
b_c	flange width, compression	9 in
b_t	flange width, tension	9 in
D	clear distance between flanges	22.54 in
F_{yc}	specified minimum yield strength of the compression flange	60 kips/in2
F_{yt}	specified minimum yield strength of the tension flange	60 kips/in2
F_{yw}	specified minimum yield strength of the web	60 kips/in2
P	total nominal compressive force in the concrete deck for the design of the shear connectors at the strength limit state	kips
$t_c = t_f$	flange thickness, compression	0.68 in
t_s	thickness of the slab	8 in
$t_t = t_f$	flange thickness, tension flange	0.68 in
t_w	web thickness	0.44 in
C_{rt}	distance from top of concrete deck to top layer of longitudinal concrete deck reinforcement	3 in
C_{rb}	distance from top of concrete deck to bottom layer of longitudinal concrete deck reinforcement	5 in

Locate the plastic neutral axis (PNA).

A App. Tbl. D6.1-1

The plastic compressive force in the slab is

$$P_s = 0.85\, f'_c b_s t_s$$

$$= (0.85)(4.5\ \text{kips/in}^2)(87\ \text{in})(8\ \text{in})$$

$$= 2662.2\ \text{kips}$$

Neglect strength of slab reinforcement.

$$P_{rb} = P_{rt} = 0 \text{ [conservative]}$$

The plastic force in the compression flange is

$$P_c = F_{yc} b_c t_c = (60\ \text{ksi})(9\ \text{in})(0.68\ \text{in}) = 367.2\ \text{kips}$$

The plastic force in the web is

$$P_w = F_{yw} D t_w = (60\ \text{ksi})(22.54\ \text{in})(0.44\ \text{in}) = 595\ \text{kips}$$

The plastic force in the tension flange is

$$P_t = F_{yt} b_t t_t = (60\ \text{ksi})(9\ \text{in})(0.68\ \text{in}) = 367.2\ \text{kips}$$

Case I (in web): $P_t + P_w \geq P_c + P_s$

A App. Tbl. D6.1-1

$$P_t + P_w = 367.2\ \text{kips} + 595\ \text{kips} = 962.2\ \text{kips}$$

$$P_c + P_s = 367.2\ \text{kips} + 2662.2\ \text{kips} = 3029.4\ \text{kips}$$

$$962.2\ \text{kips} < 3029.4\ \text{kips [no good]}$$

Case II (in top flange): $P_t + P_w + P_c \geq P_s$

$$P_t + P_w + P_c = 367.2\ \text{kips} + 595\ \text{kips} + 367.2\ \text{kips} = 1329.4\ \text{kips}$$

$$P_s = 2662.2\ \text{kips}$$

$$1329.4\ \text{kips} < 2662.2\ \text{kips [no good]}$$

Case III (in deck, below bottom reinforcement): $P_t + P_w + P_c \geq (C_{rb}/t_s)P_s$

$$P_t + P_w + P_c = 367.2\ \text{kips} + 595\ \text{kips} + 367.2\ \text{kips} = 1329.4\ \text{kips}$$

$$(C_{rb}/t_s)P_s = (5\ \text{in}/8\ \text{in})(2662.2) = 1663.9\ \text{kips}$$

$$1329.4\ \text{kips} < 1663.9\ \text{kips [no good]}$$

Case IV (not considered because the effects of slab reinforcement are ignored: $P_{rb} = 0$ and $P_{rt} = 0$)

Case V (in deck, between reinforcement layers): $P_t + P_w + P_c \geq (C_{rt}/t_s)P_s$

$$P_t + P_w + P_c = 367.2 \text{ kips} + 595 \text{ kips} + 367.2 \text{ kips} = 1329.4 \text{ kips}$$

$$(C_{rt}/t_s)P_s = (3 \text{ in}/8 \text{ in})(2937.6) = 998.3 \text{ kips}$$

$$1329.4 \text{ kips} > 998.3 \text{ kips [OK]}$$

The PNA is between the top and bottom layers of reinforcement. The PNA, $\bar{Y}$, for Case V is

A App. Tbl. D6.1-1

$$\bar{Y} = t_s\left(\frac{P_{rb} + P_c + P_w + P_t - P_{rt}}{P_s}\right)$$

$$= (8 \text{ in})\left(\frac{0 \text{ kips} + 367.2 \text{ kips} + 595 \text{ kips} + 367.2 \text{ kips} - 0 \text{ kips}}{2662.2 \text{ kips}}\right)$$

$$= 4.0 \text{ in from top of deck (was 3.62 in for interior girder)}$$

The distance from the mid-depth of steel compression flange to the PNA is

$$d_c = 8 \text{ in} + 2 \text{ in} + (0.68 \text{ in})/2 - 4.0 \text{ in} = 6.34 \text{ in}$$

The distance from the mid-depth of steel web to the PNA is

$$d_w = 8 \text{ in} + 2 \text{ in} + (23.9 \text{ in})/2 - 4.0 \text{ in} = 17.95 \text{ in}$$

The distance from the mid-depth of steel tension flange to the PNA is

$$d_t = 8 \text{ in} + 2 \text{ in} + 23.9 \text{ in} - (0.68 \text{ in})/2 - 4.0 \text{ in} = 29.65 \text{ in}$$

See Figure 2.35.

The plastic moment capacity, M_p, for Case V in exterior girders is

A Tbl. D6.1-1

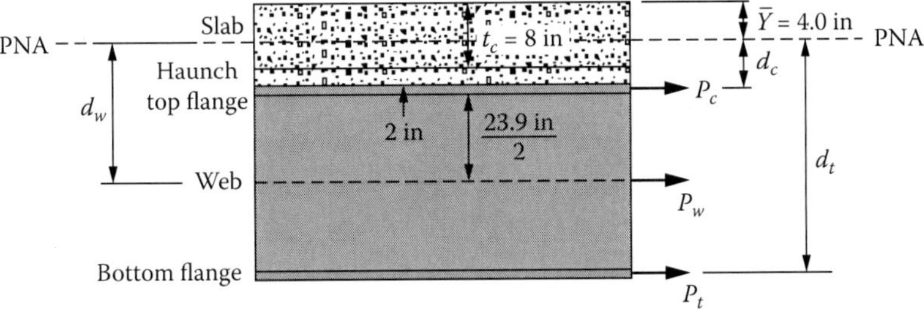

FIGURE 2.35
Section and cross section of exterior girder for plastic moment capacity.

$$M_p = \left(\frac{\bar{Y}^2 P_s}{2t_s}\right) + \left(P_{rt}d_{rt} + P_{rb}d_{rb} + P_c d_c + P_w d_w + P_t d_t\right)$$

$$= \left(\frac{(4.0\ \text{in})^2(2662.2\ \text{kips})}{2(8\ \text{in})}\right) + \left(\begin{array}{l} 0+0+(367.2\ \text{kips})(6.34\ \text{in}) \\ +(595\ \text{kips})(17.95\ \text{in}) \\ +(367.2\ \text{kips})(29.56\ \text{in}) \end{array}\right)$$

$$= 2796.8\ \text{in-kips} + 2328.0\ \text{in-kips} + 10680.3\ \text{in-kips} + 10854.4\ \text{in-kips}$$

$$= 26{,}659.5\ \text{in-kips}(1\ \text{ft}/12\ \text{in})$$

$$= 2221.6\ \text{ft-kips}$$

Assume an unstiffened web.

A Art. 6.10.9.2

NOTE: Neglect strength of slab reinforcement ($P_{rt} = P_{rb} = 0$).

The plastic shear resistance of the webs for both interior and exterior girders is

A Eq. 6.10.9.2-2

$$V_p = (0.58)F_{yw}Dt_w$$

$$= (0.58)(60\ \text{ksi})(22.54\ \text{in})(0.44\ \text{in})$$

$$= 345.2\ \text{kips}$$

TABLE 2.7

Summary of Plastic Moment Capacity and Shear Force

Girder Location	M_p, ft-kips	V_p, kips
Interior	2286	345.2
Exterior	2221	345.2

Please see Table 2.7.

Check the LRFD equation for the Strength I Limit State.

A Art. 6.10.6.2

Check the flexure for interior girders.

Confirm that the section is compact in positive flexure.

A Art. 6.10.6.2.2

$$F_y = 60 \text{ ksi} < 70 \text{ ksi [OK]}$$
$$\frac{D}{t_w} = \frac{22.54 \text{ in}}{0.44 \text{ in}} = 51.2 < 150 \text{ [OK]}$$

A Art. 6.10.2.1.1

The depth of the girder web in compression at the plastic moment, D_{cp}, is zero because the PNA is in the slab.

A App. D6.3.2

The compact composite sections shall satisfy:

Art. 6.10.6.2.2; Eq. 6.10.6.2.2-1

$$\frac{2D_{cp}}{t_w} = \frac{(2)(0)}{0.44 \text{ in}} = 0 \le 3.76\sqrt{\frac{E}{F_{yc}}} \text{ [OK]}$$

The section qualifies as a compact composite section.

The distance from the top of concrete deck to the neutral axis of composite section, D_p $(=\bar{Y})$, is

$$D_p = 3.62 \text{ in for interior girders}$$

$$D_p = 4.0 \text{ in for exterior girders}$$

The total depth of composite section, D_t, is

$$D_t = 23.9 \text{ in} + 2 \text{ in} + 8 \text{ in} = 33.9 \text{ in}$$

$$0.1\ D_t = (0.1)(33.9 \text{ in}) = 3.39 \text{ in}$$

Determine the nominal flexural resistance for interior girders, M_n.

A Art. 6.10.7.1.2

If $D_p \le 0.1\ D_t$, then $M_n = M_p$. Because $D_p = 3.62 \text{ in} \ge 0.1\ D_t = 3.39$ in, the nominal flexural resistance is

A Eq. 6.10.7.1.2-2

$$M_n = M_p\left(1.07 - 0.7\frac{D_p}{D_t}\right)$$

$$= (2286 \text{ ft-kips})\left(1.07 - 0.7\left(\frac{3.62 \text{ in}}{33.9 \text{ in}}\right)\right)$$

$$= 2275.1 \text{ ft-kips}$$

The flange lateral bending stress, f_l, is assumed to be negligible.

A Comm. 6.10.1

The resistance factor for flexure, Φ_f, is 1.0.

A Art. 6.5.4.2

Check the following.

At the strength limit state, the section shall satisfy:

A Art. 6.10.7.1, A Eq. 6.10.7.1-1

$$M_u + \frac{1}{3} f_l S_{xt} \le \varphi_f M_n = 1167 \text{ ft-kips} + \frac{1}{3}(0) S_{xt}$$

$$= 1167 \text{ ft-kips}$$

$$\varphi_f M_n = (1.0)(2275.1 \text{ ft-kips})$$

$$= 2275.1 \text{ ft-kips} \ge 1167 \text{ ft-kips [OK]}$$

Check the flexure for exterior girders.

Confirm that the section is a compact composite section.

It is the same as the interior girders; therefore, the section qualifies as a compact section.

The distance from the top of concrete deck to the neutral axis of composite section, D_p (= $\bar{Y}$), is 4.0 in.

The total depth of composite section, D_t, is 33.9 in.

Determine the nominal flexural resistance for exterior girders, M_n

A Art. 6.10.7.1.2

If $D_p \le D_t$, then $M_n = M_p$. Because $D_p = 4.0$ in > 0.1 $D_t = 3.39$ in; therefore, the nominal flexural resistance is

A Eq. 6.10.7.1.2-2

$$M_n = M_p\left(1.07 - 0.7\frac{D_p}{D_t}\right)$$

$$= (2264 \text{ ft-kips})\left(1.07 - 0.7\left(\frac{4.0 \text{ in}}{33.9 \text{ in}}\right)\right)$$

$$= 2235.5 \text{ ft-kips}$$

The flange lateral bending stress, f_l, is assumed to be negligible.

A Comm. 6.10.1

The resistance factor for flexure, Φ_f, is 1.0.

A Art. 6.5.4.2

Check the following.

At the strength limit state, the section shall satisfy:

A Art. 6.10.7.1, A Eq. 6.10.7.1-1

$$M_u + \frac{1}{3}f_l S_{xt} \le \varphi_f M_n = 1288.0 \text{ ft-kips} + \frac{1}{3}(0)S_{xt}$$

$$= 1288.0 \text{ ft-kips}$$

$$\varphi_f M_n = (1.0)(2235.5 \text{ ft-kips})$$

$$= 2235.5 \text{ ft-kips} \ge 1288.0 \text{ ft-kips [OK]}$$

Find the shear resistance.

The resistance factor for shear, Φ_v, is 1.0.

A Art. 6.5.4.2

The nominal shear resistance of the unstiffened webs is

A Art. 6.10.9.2; A Eq. 6.10.9.2-1

$$V_n = V_{cr} = CV_p$$

The ratio of shear-buckling resistance to shear yield strength, C, is 1.0 if the following is true.

A Art. 6.10.9.3.2; Eq. 6.10.9.3.2-4

$$\frac{D}{t_w} \le 1.12\sqrt{\frac{Ek}{F_{yw}}}$$

$$\frac{D}{t_w} = \frac{22.54 \text{ in}}{0.44 \text{ in}} = 51.2$$

The shear-buckling coefficient, k, is

A Eq. 6.10.9.3.2-7

$$k = 5 + \frac{S}{\left(\frac{d_o}{D}\right)^2}$$

which is assumed to be equal to 5.0 inasmuch as d_o, the transverse spacing, is large, i.e., no stiffeners.

$$1.12\sqrt{\frac{Ek}{F_{yw}}} = 1.12\sqrt{\frac{(29{,}000 \text{ ksi})(5)}{60 \text{ ksi}}} = 55.1$$

Because $(D/t_w) = 51.2 < 55.1$, C is 1.0.

$$V_n = CV_p = (1.0)V_p = V_p = 345.2 \text{ kips}$$

A Eq. 6.10.9.2-1

Check that the requirement is satisfied for interior girders.

A Eq. 6.10.9.1-1; Art. 6.10.9.1

$$V_u \leq \Phi_v V_n$$

$$156.3 \text{ kips} < (1.0)(345.2 \text{ kips}) = 345.2 \text{ kips [OK]}$$

Check that the requirement is satisfied for exterior girders.

$$V_u \leq \Phi_v V_n$$

$$146.7 \text{ kips} < (1.0)(345.2 \text{ kips}) = 345.2 \text{ kips [OK]}$$

Check the Service II Limit State LRFD equation for permanent deformations.

A Tbls. 3.4.1-1, 3.4.1-2, 6.10.4.2

$$Q = 1.0(DC) + 1.0(DW) + 1.30(LL + IM)$$

Calculate the flange stresses, f_f, for interior girders due to Service II loads.

The top steel flange for the composite section due to Service II loads must satisfy the following:

A Eq. 6.10.4.2.2-1

$$f_f \leq 0.95\, R_h F_{yf}$$

The hybrid factor, R_h, is 1.0.

A Art. 6.10.1.10.1

The yield strength of the flange, F_{yf}, is 60 ksi (given).

Transform the concrete deck to an equivalent steel area with modular ratio between steel and concrete, n, of 8. Please see Figures 2.36, 2.37, and 2.38.

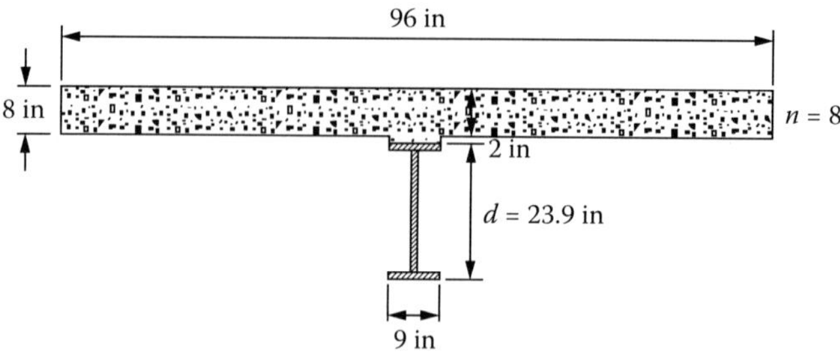

FIGURE 2.36
Interior girder section prior to transformed area.

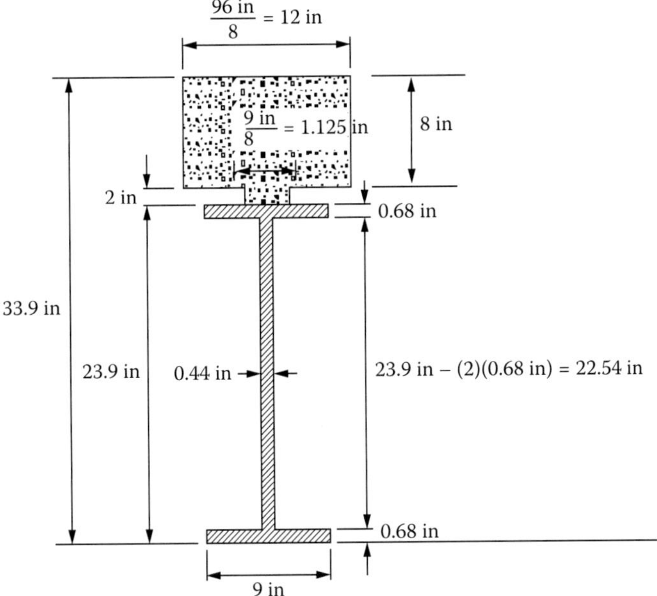

FIGURE 2.37
Interior girder section after transformed area.

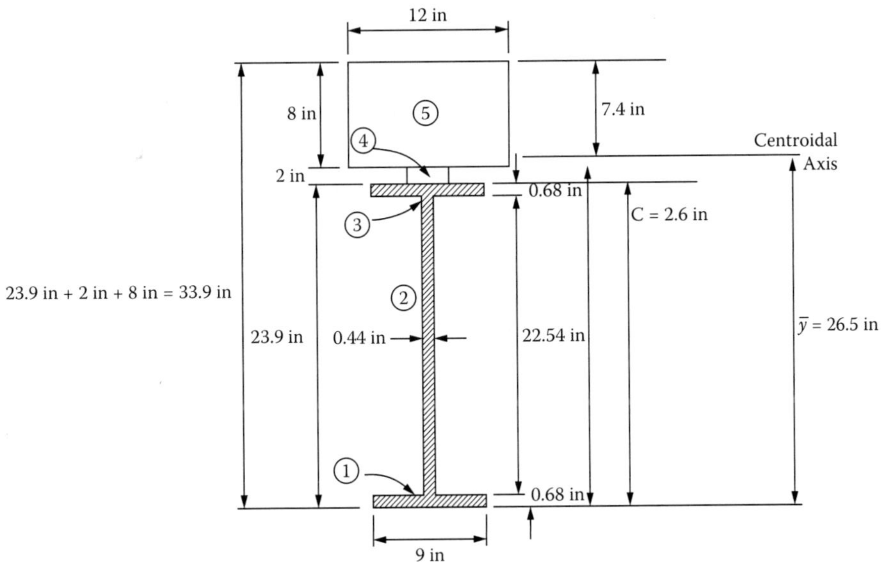

FIGURE 2.38
Dimensions for transformed interior beam section.

Calculate the dimensions for the transformed interior girder section.

$$A_1 = (0.68\text{ in})(9.0\text{ in}) = 6.12\text{ in}^2$$

$$A_2 = (0.44\text{ in})(22.45\text{ in}) = 9.92\text{ in}^2$$

$$A_3 = (0.68\text{ in})(9.0\text{ in}) = 6.12\text{ in}^2$$

$$A_4 = (1.125\text{ in})(2.0\text{ in}) = 2.25\text{ in}^2$$

$$A_5 = (12\text{ in})(8.0\text{ in}) = 96.0\text{ in}^2$$

$$\Sigma A = A_1 + A_2 + A_3 + A_4 + A_5$$

$$= 6.12\text{ in}^2 + 9.92\text{ in}^2 + 6.12\text{ in}^2 + 2.25\text{ in}^2 + 96\text{ in}^2$$

$$= 120.4\text{ in}^2$$

Calculate the products of A and $\bar{Y}$ (from beam bottom).

$$A_1\bar{Y}_1 = \left(6.12\text{ in}^2\right)\left(\frac{0.68\text{ in}}{2}\right) = 2.08\text{ in}^3$$

$$A_2\bar{Y}_2 = \left(9.92\text{ in}^2\right)\left(\frac{22.54\text{ in}}{2}\right) = 118.54\text{ in}^3$$

$$A_3\bar{Y}_3 = \left(6.12\text{ in}^2\right)\left(23.9\text{ in} - \frac{0.68\text{ in}}{2}\right) = 144.2\text{ in}^3$$

$$A_4\bar{Y}_4 = \left(2.25\text{ in}^2\right)\left(23.9\text{ in} - \frac{2\text{ in}}{2}\right) = 56.03\text{ in}^3$$

$$A_5\bar{Y}_5 = \left(96\text{ in}^2\right)\left(23.9\text{ in} + 2\text{ in} + \frac{8\text{ in}}{2}\right) = 2870.4\text{ in}^3$$

$$\Sigma AY = A_1\bar{Y}_1 + A_2\bar{Y}_2 + A_3\bar{Y}_3 + A_4\bar{Y}_4 + A_5\bar{Y}_5$$

$$= 2.08\text{ in}^3 + 118.54\text{ in}^3 + 144.2\text{ in}^3 + 56.03\text{ in}^3 + 2870.4\text{ in}^3$$

$$= 3191.5\text{ in}^3$$

The centroid of transformed section from the bottom of beam, 5, is

$$\bar{y} = \frac{\Sigma A\bar{y}}{\Sigma A} = \frac{3191.5\text{ in}^3}{120.4\text{ in}^2} = 26.5\text{ in}$$

from bottom of beam to the centroidal axis.

The moments of inertia are calculated as follows.

$$I = \frac{bh^3}{12} + Ad^2$$

$$I_1 = \frac{(9\text{ in})(0.68\text{ in})^3}{12} + (6.12\text{ in}^2)\left(26.5\text{ in} - \frac{0.68\text{ in}}{2}\right)^2 = 4188.4\text{ in}^4$$

$$I_2 = \frac{(0.44\text{ in})(22.54\text{ in})^3}{12} + (9.92\text{ in}^2)\left(26.5\text{ in} - 0.68\text{ in} - \frac{22.54\text{ in}}{2}\right)^2 = 2520.0\text{ in}^4$$

$$I_3 = \frac{(9\text{ in})(0.68\text{ in})^3}{12} + (6.12\text{ in}^2)\left(26.5\text{ in} - 0.68\text{ in} - 22.54\text{ in} - \frac{0.68\text{ in}}{2}\right)^2 = 53.1\text{ in}^4$$

$$I_4 = \frac{(1.125\text{ in})(2\text{ in})^3}{12} + (2.25\text{ in}^2)(26.5\text{ in} - 23.9\text{ in} - 1\text{ in})^2 = 6.5\text{ in}^4$$

$$I_5 = \frac{(12\text{ in})(8\text{ in})^3}{12} + (96\text{ in}^2)(7.6\text{ in} - 4\text{ in})^2 = 1756.2\text{ in}^4$$

$$I_{total} = I_1 + I_2 + I_3 + I_4 + I_5$$

$$= 4188.4\text{ in}^4 + 2520.0\text{ in}^4 + 53.1\text{ in}^4 + 6.5\text{ in}^4 + 1756.2\text{ in}^4$$

$$= 8524.2\text{ in}^4$$

The distance between $\bar{y}$ and the top of the top flange of the steel section is

$$c = 26.5\text{ in} - 23.9\text{ in} = 2.6\text{ in}$$

The section modulus at the steel top flange is

$$S_{top} = \frac{I}{c} = \frac{8524.2\text{ in}^4}{2.6\text{ in}} = 3278.5\text{ in}^3$$

The moment for composite dead load per interior or exterior girder, M_{DC}, is 223.6 ft-kips; the moment for wearing surface per girder, M_{DW}, is 36.6 ft-kips; and the total moment per interior or exterior girder, M_{LL+IM}, is 475.9 ft-kips.

The steel top flange stresses, f_f, due to Service II loads are

$$f_{DC} = \frac{M_{DC}}{S_{top}} = \frac{(223.6\ \text{ft-kips})\left(12\frac{\text{in}}{\text{ft}}\right)}{3278.5\ \text{in}^3} = 0.82\frac{\text{kips}}{\text{in}^2}$$

$$f_{DW} = \frac{M_{DW}}{S_{top}} = \frac{(36.6\ \text{ft-kips})\left(12\frac{\text{in}}{\text{ft}}\right)}{3278.5\ \text{in}^3} = 0.13\frac{\text{kips}}{\text{in}^2}$$

$$f_{LL+IM} = \frac{M_{LL+IM}}{S_{top}} = \frac{(475.9\ \text{ft-kips})\left(12\frac{\text{in}}{\text{ft}}\right)}{3278.5\ \text{in}^3} = 1.74\frac{\text{kips}}{\text{in}^2}$$

$f_f = 1.0\ f_{DC} + 1.0\ f_{DW} + 1.3\ f_{LL+IM}$

A Tbl. 3.4.1-1

$= (1.0)(0.82\ \text{kips/in}^2) + (1.0)(0.13\ \text{kips/in}^2) + (1.3)(1.74\ \text{kips/in}^2)$

$= 3.21\ \text{kips/in}^2$

Confirm that the steel top flange of the composite section due to Service II loads satisfies the requirement.

A Art. 6.10.4.2, A Eq. 6.10.4.2.2-1

$f_f \le 0.95\ R_h F_{yf}$

The hybrid factor, R_h, is 1.0.

A Art. 6.10.1.10.1

$0.95\ R_h F_{yf} = (0.95)(1.0)(60\ \text{kips/in}^2)$

$= 57\ \text{kips/in}^2 > f_f = 3.21\ \text{kips/in}^2$ [OK]

Calculate the bottom steel flange stresses for composite sections, satisfying the following.

The bottom flange shall satisfy

A Art. 6.10.4.2, A Eq. 6.10.4.2.2-2

$$f_f + \frac{f_l}{2} \le 0.95\ R_h F_{yf}$$

The flange lateral bending stress, f_l, is negligible (0.0).

The distance from the extreme tension fiber to the neutral axis is $c = \bar{Y} =$ 26.5 in.

The section modulus of the bottom flange is

$$S_{bottom} = \frac{I}{c} = \frac{8524.2 \text{ in}^4}{26.5 \text{ in}} = 321.7 \text{ in}^3$$

The bottom flange steel stresses, f_f, due to Service II loads are

$$f_{DC} = \frac{M_{DC}}{S_{bottom}} = \frac{(223.6 \text{ ft-kips})\left(12 \frac{\text{in}}{\text{ft}}\right)}{321.7 \text{ in}^3} = 8.34 \text{ kips/in}^2$$

$$f_{DW} = \frac{M_{DW}}{S_{bottom}} = \frac{(36.6 \text{ ft-kips})\left(12 \frac{\text{in}}{\text{ft}}\right)}{321.7 \text{ in}^3} = 1.36 \text{ kips/in}^2$$

$$f_{LL+IM} = \frac{M_{LL+IM}}{S_{bottom}} = \frac{(475.9 \text{ ft-kips})\left(12 \frac{\text{in}}{\text{ft}}\right)}{321.7 \text{ in}^3} = 17.75 \text{ kips/in}^2$$

$f_f = 1.0\, f_{DC} + 1.0\, f_{DW} + 1.3\, f_{LL+IM}$

A Tbl. 3.4.1-1

$= (1.0)(8.34 \text{ kips/in}^2) + (1.0)(1.36 \text{ kips/in}^2) + (1.3)(17.75 \text{ kips/in}^2)$

$= 32.78 \text{ kips/in}^2$

Confirm that the steel bottom flange of composite section due to Service II loads satisfies the requirement.

A Art. 6.10.4.2, A Eq. 6.10.4.2.2-2

$$f_f + \frac{f_l}{2} \le 0.95\, R_h F_{yf}$$

$$32.78 \frac{\text{kips}}{\text{in}^2} + \frac{0 \frac{\text{kips}}{\text{in}^2}}{2} \le 0.95(1.0)\left(60 \frac{\text{kips}}{\text{in}^2}\right)$$

$32.78 \text{ kips/in}^2 < 57 \text{ kips/in}^2$ [OK]

Calculate the flange stresses for exterior girders. Please see Figure 2.39.

A Art. 6.10.4.2

Transform the concrete deck area to an equivalent area of steel using a modular ratio between steel and concrete, n, of 8. Please see Figures 2.40 and 2.41.

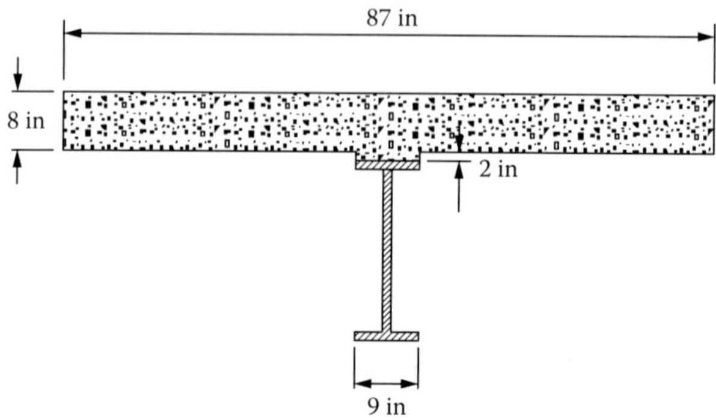

FIGURE 2.39
Exterior girder section prior to transformed area.

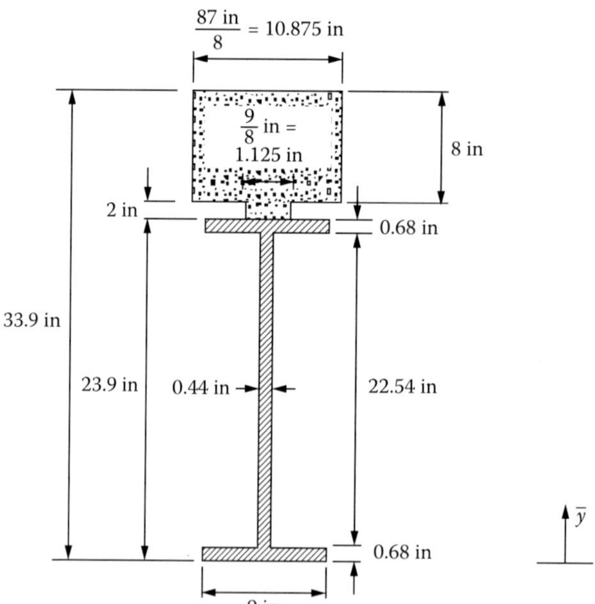

FIGURE 2.40
Exterior girder section after transformed area.

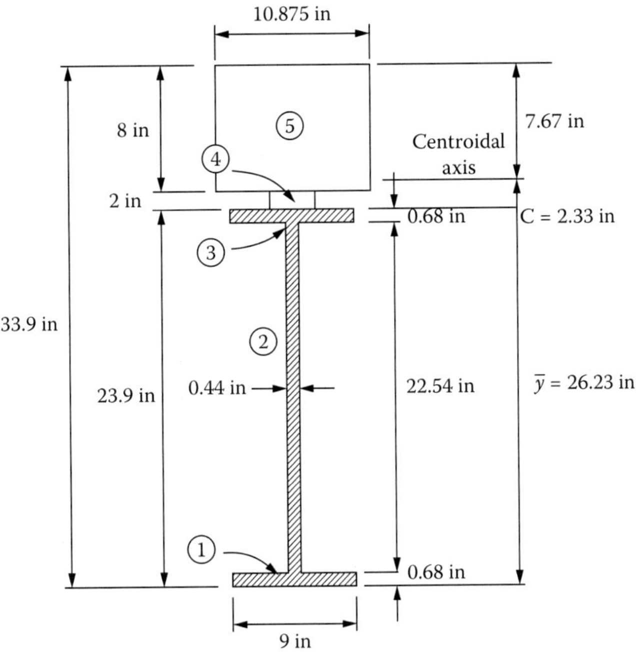

FIGURE 2.41
Dimensions for transformed exterior beam section.

Calculate the dimensions for the transformed exterior girder section.

$$A_1 = (0.68 \text{ in})(9.0 \text{ in}) = 6.12 \text{ in}^2$$

$$A_2 = (0.44 \text{ in})(22.45 \text{ in}) = 9.92 \text{ in}^2$$

$$A_3 = (0.68 \text{ in})(9.0 \text{ in}) = 6.12 \text{ in}^2$$

$$A_4 = (1.125 \text{ in})(2.0 \text{ in}) = 2.25 \text{ in}^2$$

$$A_5 = (10.875 \text{ in})(8.0 \text{ in}) = 87.0 \text{ in}^2$$

$$\Sigma A = A_1 + A_2 + A_3 + A_4 + A_5$$

$$= 6.12 \text{ in}^2 + 9.92 \text{ in}^2 + 6.12 \text{ in}^2 + 2.25 \text{ in}^2 + 87 \text{ in}^2$$

$$= 111.4 \text{ in}^2$$

Calculate the products of A and $\bar{Y}$ (from beam bottom).

$$A_1\bar{Y}_1 = (6.12\text{ in}^2)\left(\frac{0.68\text{ in}}{2}\right) = 2.08\text{ in}^3$$

$$A_2\bar{Y}_2 = (9.92\text{ in}^2)\left(\frac{22.54\text{ in}}{2}\right) = 118.54\text{ in}^3$$

$$A_3\bar{Y}_3 = (6.12\text{ in}^2)\left(23.9\text{ in} - \frac{0.68\text{ in}}{2}\right) = 144.2\text{ in}^3$$

$$A_4\bar{Y}_4 = (2.25\text{ in}^2)\left(23.9\text{ in} - \frac{2\text{ in}}{2}\right) = 56.03\text{ in}^3$$

$$A_5\bar{Y}_5 = (87\text{ in}^2)\left(23.9\text{ in} + 2\text{ in} + \frac{8\text{ in}}{2}\right) = 2601.3\text{ in}^3$$

$$\Sigma AY = A_1\bar{Y}_1 + A_2\bar{Y}_2 + A_3\bar{Y}_3 + A_4\bar{Y}_4 + A_5\bar{Y}_5$$

$$= 2.08\text{ in}^3 + 118.54\text{ in}^3 + 144.2\text{ in}^3 + 56.03\text{ in}^3 + 2601.3\text{ in}^3$$

$$= 2922.15\text{ in}^3$$

The centroid of transformed section from the bottom of beam, $\bar{y}$, is

$$\bar{y} = \frac{\Sigma A\bar{y}}{\Sigma A} = \frac{2922.15\text{ in}^3}{111.4\text{ in}^2} = 26.23\text{ in} \text{ from bottom of beam}$$

The moments of inertia are calculated as follows.

$$I = \frac{bh^3}{12} + Ad^2$$

$$I_1 = \frac{(9\text{ in})(0.68\text{ in})^3}{12} + (6.12\text{ in}^2)\left(26.23\text{ in} - \frac{0.68\text{ in}}{2}\right)^2 = 4102.4\text{ in}^4$$

$$I_2 = \frac{(0.44\text{ in})(22.54\text{ in})^3}{12} + (9.92\text{ in}^2)\left(26.23\text{ in} - 0.68\text{ in} - \frac{22.54\text{ in}}{2}\right)^2 = 2442.8\text{ in}^4$$

$$I_3 = \frac{(9\text{ in})(0.68\text{ in})^3}{12} + (6.12\text{ in}^2)\left(26.23\text{ in} - 0.68\text{ in} - 22.54\text{ in} - \frac{0.68\text{ in}}{2}\right)^2 = 43.9\text{ in}^4$$

$$I_4 = \frac{(1.125\text{ in})(2\text{ in})^3}{12} + (2.25\text{ in}^2)(26.23\text{ in} - 23.9\text{ in} - 1\text{ in})^2 = 4.7\text{ in}^4$$

$$I_5 = \frac{(10.875\text{ in})(8\text{ in})^3}{12} + (87\text{ in}^2)(8.67\text{ in} - 4\text{ in})^2 = 1635.8\text{ in}^4$$

$$I_{total} = I_1 + I_2 + I_3 + I_4 + I_5$$

$$= 4102.4\text{ in}^4 + 2442.8\text{ in}^4 + 43.9\text{ in}^4 + 4.7\text{ in}^4 + 1635.8\text{ in}^4 = 8230.0\text{ in}^4$$

Calculate the top steel flange stresses for composite sections, satisfying the following:

$$f_f \leq 0.95\, R_h F_{yf}$$

A Eq. 6.10.4.2-1

The hybrid factor, R_h, is 1.0.

The distance between $\bar{y}$ and the top of the top flange of the steel section is

$$c = 26.23\text{ in} - 23.9\text{ in} = 2.33\text{ in}$$

The section modulus for the top flange of the steel section is

$$S_{top} = \frac{I}{c} = \frac{8230.0\text{ in}^4}{2.33\text{ in}} = 3532.2\text{ in}^3$$

The steel top flange stresses, f_f, due to Service II loads are

$$f_{DC} = \frac{M_{DC}}{S_{top}} = \frac{(223.6\text{ ft-kips})\left(12\frac{\text{in}}{\text{ft}}\right)}{3532.2\text{ in}^3} = 0.76\frac{\text{kips}}{\text{in}^2}$$

$$f_{DW} = \frac{M_{DW}}{S_{top}} = \frac{(36.6\text{ ft-kips})\left(12\frac{\text{in}}{\text{ft}}\right)}{3532.2\text{ in}^3} = 0.12\frac{\text{kips}}{\text{in}^2}$$

$$f_{LL+IM} = \frac{M_{LL+IM}}{S_{top}} = \frac{(544.9\text{ ft-kips})\left(12\frac{\text{in}}{\text{ft}}\right)}{3532.2\text{ in}^3} = 1.85\frac{\text{kips}}{\text{in}^2}$$

$$f_f = 1.0\ f_{DC} + 1.0\ f_{DW} + 1.3\ f_{LL+IM}$$

A Tbl. 3.4.1-1

$$= (1.0)(0.76\text{ kips/in}^2) + (1.0)(0.12\text{ kips/in}^2) + (1.3)(1.85\text{ kips/in}^2)$$

$$= 3.29\text{ kips/in}^2$$

Confirm that the steel top flange of composite section due to Service II loads satisfies the requirement.

A Art. 6.10.4.2

$$f_f \le 0.95\ R_h F_{yf}$$

A Eq. 6.10.4.2.2-1

$$0.95\ R_h F_{yf} = (0.95)(1.0)(60\text{ kips/in}^2)$$

$$= 57\text{ kips/in}^2 > f_f = 3.29\text{ kips/in}^2\ \text{[OK]}$$

Calculate the bottom steel flange stresses for composite sections due to Service II loads satisfying the following.

Flange shall satisfy,

A Art. 6.10.4.2. A Eq. 6.10.4.2.2-2

$$f_f + \frac{f_l}{2} \le 0.95\ R_h F_{yf}$$

The flange lateral bending stress, f_l, is negligible (0.0).

The hybrid factor, R_h, is 1.0.

A Art. 6.10.1.10.1

The distance from the extreme tension fiber to the neutral axis is $c = \bar{y} = 26.23$ in.

The section modulus of the bottom flange is

$$S_{bottom} = \frac{I}{c} = \frac{8230.0\text{ in}^4}{26.23\text{ in}} = 313.76\text{ in}^3$$

The bottom flange steel stresses, f_f, due to Service II loads are

$$f_{DC} = \frac{M_{DC}}{S_{bottom}} = \frac{(223.6 \text{ ft-kips})\left(12\frac{\text{in}}{\text{ft}}\right)}{313.76 \text{ in}^3} = 8.55 \text{ kips/in}^2$$

$$f_{DW} = \frac{M_{DW}}{S_{bottom}} = \frac{(36.6 \text{ ft-kips})\left(12\frac{\text{in}}{\text{ft}}\right)}{313.76 \text{ in}^3} = 1.40 \text{ kips/in}^2$$

$$f_{LL+IM} = \frac{M_{LL+IM}}{S_{bottom}} = \frac{(544.9 \text{ ft-kips})\left(12\frac{\text{in}}{\text{ft}}\right)}{313.76 \text{ in}^3} = 20.84 \text{ kips/in}^2$$

$f_f = 1.0\ f_{DC} + 1.0\ f_{DW} + 1.3\ f_{LL+IM}$

A Tbl. 3.4.1-1

$= (1.0)(8.55 \text{ kips/in}^2) + (1.0)(1.40 \text{ kips/in}^2) + (1.3)(20.84 \text{ kips/in}^2)$

$= 37.04 \text{ kips/in}^2$

Confirm that the steel bottom flange of composite section due to Service II loads satisfies the requirement.

A Art. 6.10.4.2, A Eq. 6.10.4.2.2-2

$$f_f + \frac{f_l}{2} \leq 0.95_f R_h F_{yf}$$

$$37.04\frac{\text{kips}}{\text{in}^2} + \frac{0\frac{\text{kips}}{\text{in}^2}}{2} \leq 0.95(1.0)\left(60\frac{\text{kips}}{\text{in}^2}\right)$$

$37.04 \text{ kips/in}^2 < 57 \text{ kips/in}^2$ [OK]

Check LRFD Fatigue Limit State II.

Details shall be investigated for the fatigue as specified in AASHTO Art. 6.6.1. The fatigue load combination in AASHTO Tbl. 3.4.1-1 and the fatigue load specified in AASHTO Art. 3.6.1.4 shall apply.

A Art. 6.6.1; 3.6.1.4

$$Q = 0.75(LL + IM)$$

A Tbl. 3.4.1-1

For load-induced fatigue, each detail shall satisfy,

A Eq. 6.6.1.2.2-1; Art. 6.6.1.2

$$\gamma(\Delta f) \leq (\Delta F)_n$$

The force effect, live load stress range due to fatigue load is (Δf). The load factor is γ. The nominal fatigue resistance is $(\Delta F)_n$.

A 6.6.1.2.2

The fatigue load is one design truck with a constant spacing of 30 ft between 32 kip axles.

A Art. 3.6.1.4.1

Find the moment due to fatigue load.

Please see Figure 2.42.
Determine the moment due to fatigue load.

$$\Sigma M_{@B} = 0$$

$$R_A(40\text{ ft}) = (32\text{ kips})(20\text{ ft}) + (8\text{ kips})(6\text{ ft})$$

$$R_A = 17.2\text{ kips}$$

$$M_c = (17.2\text{ kips})(20\text{ ft}) = 344\text{ ft-kips}$$

The dynamic load allowance, IM, is 15%.

A Tbl. 3.6.2.1-1

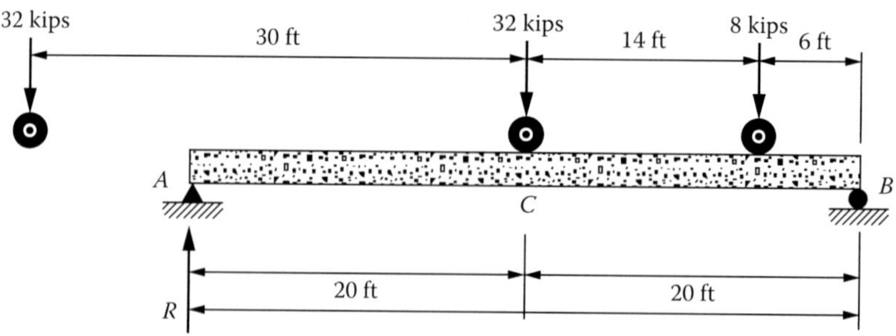

FIGURE 2.42
Single lane fatigue load placement with one design truck load for maximum moment at midspan.

The approximate load distribution factors for one traffic lane shall be used. Divide out the multiple presence factor, m, which was already included in the approximate equations for distribution factors, so it is 1.2.

Comm. 3.6.1.1.2; A Art. 3.6.1.4.3b

The load distribution factor for fatigue moment in interior girders with one design lane loaded, is

$$DFM_{fat,int}\frac{DFM_{si}}{1.2}=\frac{0.498}{1.2}=0.415$$

The load distribution factor for fatigue moment in exterior girders with one design lane loaded is

$$DFM_{fat,ext}\frac{DFM_{se}}{1.2}=\frac{0.75}{1.2}=0.625$$

The unfactored, distributed fatigue moment is

$$M_{fat}=M_{fat,LL}=(DFM_{fat})(1+IM)$$

The unfactored, distributed fatigue moment for interior girders is

$$M_{fat,int}=(344\text{ ft-kips})(0.415)(1+0.15)=164.2\text{ ft-kips}$$

The unfactored, distributed fatigue moment for exterior girders is

$$M_{fat,ext}=(344\text{ ft-kips})(0.625)(1+0.15)=247.3\text{ ft-kips}$$

Use S_{bottom} because this causes the maximum stress in either the top steel or the bottom steel flange.

The maximum stress due to fatigue load for interior girders is

$$\Delta f_{int}=\frac{M_{fat,int}}{S_{bottom}}=\frac{164.2\text{ ft-kips}\left(12\frac{\text{in}}{\text{ft}}\right)}{321.7\text{ in}^3}=6.12\frac{\text{kips}}{\text{in}^2}$$

The maximum stress due to fatigue load for exterior girders is

$$\Delta f_{ext}=\frac{M_{fat,ext}}{S_{bottom}}=\frac{247.3\text{ ft-kips}\left(12\frac{\text{in}}{\text{ft}}\right)}{313.76\text{ in}^3}=9.46\frac{\text{kips}}{\text{in}^2}$$

The girder is in Detail Category A, because it is a plain rolled member.

A Tbl. 6.6.1.2.3-1

$A = 250 \times 10^8$ kips/in^2 (constant value taken for Detail Category A)

A Tbl. 6.6.1.2.5-1

The number of cycles per truck passage, n, is 2.0 for simple span girders with span equal to and less than 40 ft.

A Tbl. 6.6.1.2.5-2

The fraction of truck traffic in a single lane, p, is 0.80 because the number of lanes available to trucks, N_L, is 3 lanes.

A Tbl. 3.6.1.4.2-1

The single lane daily truck traffic is averaged over the design life,

$$ADTT_{SL} = (p)(ADTT) = (0.80)(2500) = 2000 \text{ trucks per day}$$

A Eq. 3.6.1.4.2-1

The number of cycles of stress range for 75 years N is.

A Art. 6.6.1.2.5; Eq. 6.6.1.2.5-3

$$N = (365 \text{ days})(75 \text{ years})n(ADTT)_{SL} = (47{,}625)(2.0)(2000) = 109{,}500{,}000 \text{ cycles}$$

The constant amplitude fatigue threshold, $(\Delta F)_{TH}$, is 24.0 kips/in^2 for Detail Category A.

A Tbl. 6.6.1.2.5-3

The nominal fatigue resistance, $(\Delta F)_n$, is

A Art. 6.6.1.2.2; Eq. 6.6.1.2.5-2

$$(\Delta F)_n = \left(\frac{A}{N}\right)^{1/3}$$

$$= \left(\frac{250\text{x}10^8 \dfrac{\text{kips}}{\text{in}^2}}{109{,}500{,}000 \text{ cycles}}\right)^{1/3}$$

$$= 6.11 \text{ kips/in}^2$$

Confirm that the following formula is satisfied.

A Art. 6.6.1.2.2

$$\gamma(\Delta f) \leq (\Delta F)_n$$

where:

γ = load factor in Table 3.4.1-1 for Fatigue II load combination

(Δf) = force effect, live load stress range due to the passage of the fatigue load (Art. 3.6.1.4)

The factored live load stress due to fatigue load for interior girders is

A Tbl. 3.4.1-1

$$\gamma(\Delta f_{int}) = (0.75)(LL + IM)(\Delta f_{int}) = (0.75)(1.0)(6.12\ \text{kips/in}^2)$$

$$= 4.59\ \text{kips/in}^2 < (\Delta F)_n = 6.11\ \text{kips/in}^2\ [\text{OK}]$$

The factored live load stress due to fatigue load for exterior girders is

$$\gamma(\Delta f_{ext}) = (0.75)(LL + IM)(\Delta f_{ext}) = (0.75)(1.0)(9.46\ \text{kips/in}^2)$$

$$= 7.09\ \text{kips/in}^2 > (\Delta F)_n = 6.11\ \text{kips/in}^2\ [\text{NG}]$$

To control web-buckling and elastic flexing of the web, provisions of AASHTO Art. 6.10.5.3 shall be satisfied for distortion-induced fatigue.

A Art. 6.10.5.3

The ratio of the shear buckling resistance to the shear specified minimum yield, C, is 1.0 (see previous calculations).

A Eq. 6.10.9.3.2-4

The plastic shear force, V_p, was 345.2 kips for both interior and exterior girders.

The shear buckling resistance, V_{cr}, is

A Eq. 6.10.9.3.3-1

$$V_n = V_{cr} = CV_p = (1.0)(345.2\ \text{kips}) = 345.2\ \text{kips}$$

Find the shear due to factored fatigue load. Please see Figure 2.43.

$$\Sigma M_{@B} = 0$$

$$R_A = 32\ \text{kips} + \left(\frac{10\ \text{ft}}{40\ \text{ft}}\right)(32\ \text{kips}) = 40\ \text{kips}$$

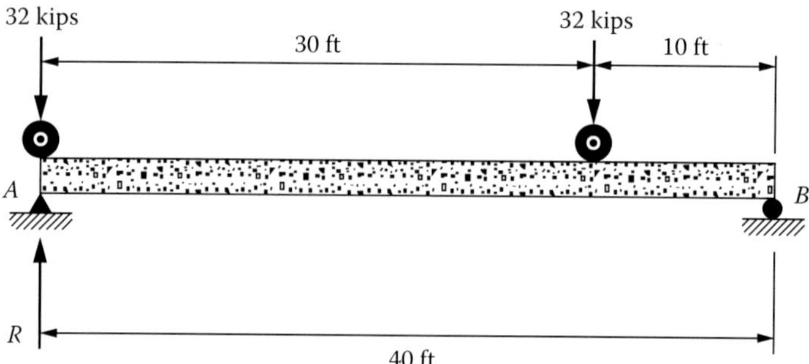

FIGURE 2.43
Single lane fatigue load placement with one design truck load for maximum shear at support.

The shear due to fatigue load, V, is 40 kips.

The distribution factor for shear due to fatigue load for interior girders with one design lane loaded is

$$DFV_{fat,int} = \frac{DFV_{int}}{1.2} = \frac{0.68}{1.2} = 0.57$$

The distribution factor for shear due to fatigue load for exterior girders with one design lane loaded is

$$DFV_{fat,ext} = \frac{DFV_{ext}}{1.2} = \frac{0.75}{1.2} = 0.625$$

The shear for the total component dead load for interior girders, V_{DC}, is 22.4 kips; and the shear for the wearing surface dead load for interior girders, V_{DW}, is 3.66 kips.

Special Fatigue Requirements for Webs

A Art. 6.10.5.3

The factored fatigue load shall be taken as twice that calculated using the fatigue load combination specified in AASHTO Tbl. 3.4.1-1, with the fatigue live load taken as in AASHTO 3.6.1.4.

Interior panels of web shall satisfy

$$V_u \leq V_{cr}$$

A Eq. 6.10.5.3-1

where:

V_u = shear in the web due to the unfactored permanent load plus the factored fatigue load

V_{cr} = shear-buckling resistance determined from AASHTO Eq. 6.10.9.3.3-1
= 345.2 kips (previously calculated)

$V_{permanent} = V_{DC} + V_{DW}$
= 22.44 kips + 3.66 kips (previously calculated)
= 26.1 kips

The factored fatigue load is

$$\gamma = \text{load factor} = 0.75 \text{ for Fatigue II load}$$

A Tbl. 3.4.1-1

$$IM = \text{dynamic allowance for fatigue} = 15\%$$

A Tbl. 3.6.2.1-1

The factored fatigue for girders is given as

$$V_f = 2(DFV)(\gamma)(V)(LL + IM)$$

For interior beams the factored fatigue load is

$$V_{f,int} = (2)(DFV_{int})(\gamma)(V)(LL + IM)$$
$$= (2)(0.57)(0.75)(40.0 \text{ kips})(1 + 0.15)$$
$$= 39.33 \text{ kips for interior beams}$$

For exterior beams the factored fatigue load is

$$V_{f,ext} = (2)(DFV_{ext})(\gamma)(V)(LL + IM)$$
$$= (2)(0.625)(0.75)(40.0 \text{ kips})(1 + 0.15)$$
$$= 43.13 \text{ kips for exterior beams [controls]}$$
$$V_u = V_{permanent} + V_f$$
$$= 26.1 \text{ kips} + 43.13 \text{ kips} = 69.23 \text{ kips}$$

Interior panel webs shall satisfy:

$$V_u \leq V_{cr}$$

A Art. 6.10.5.3; Eq. 6.10.5.3-1

$$V_u = 69.23 \text{ kips} \leq V_{cr} = 345.2 \text{ kips [OK]}$$

It is noted that the application of Art. 6.10.5.3 for Fatigue I load is considered conservative for Fatigue II load.

Design Example 4: Longitudinal Steel Girder

Situation

A simple span noncomposite steel girder rural highway bridge with a span of 40 ft is shown in Figure 2.44.

The overall width of the bridge is 33 ft 4 in.

The clear (roadway) width is 28 ft 0 in.

The roadway is a concrete slab 7 in thick (dead load of 145 lbf/ft^3, $f_c' = 4$ kips/in^2) supported by four W33 x 130, A992 Grade 50 steel girders that are spaced at 8 ft 4 in apart.

The compression flange is continuously supported by the concrete slab, and additional bracing is provided at the ends and at midspan.

Noncomposite construction is assumed.

There is a wearing surface of 3 in thick bituminous pavement (dead load of 0.140 kips/ft^3).

The barrier, sidewalk, and railings combine for a dead load of 1 kip/ft.

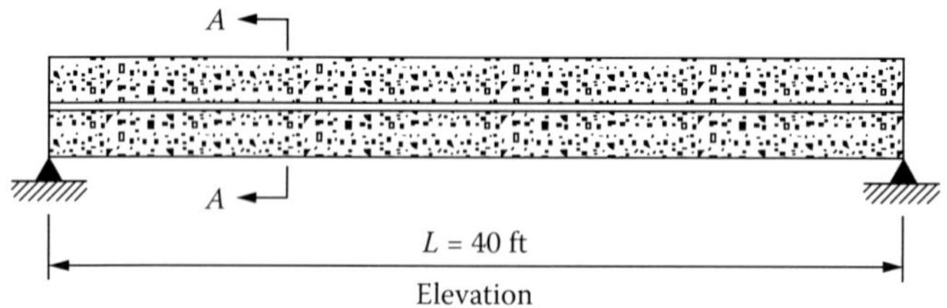

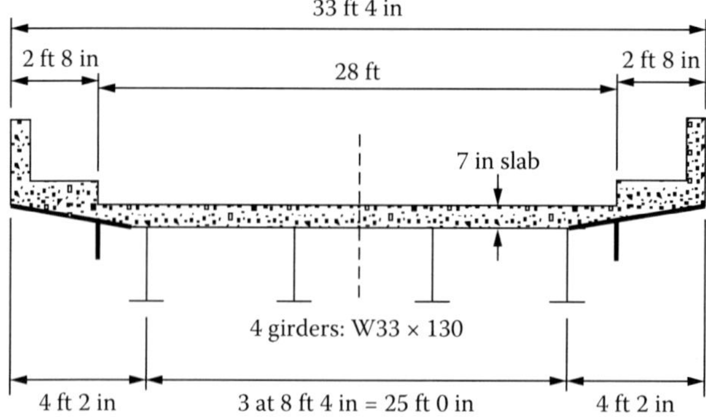

FIGURE 2.44
Steel girder bridge, 40 ft span.

Requirements

Review the longitudinal girders for adequacy against maximum live load, dead load, and shear load for load combination limit states Strength I, Service I, and Fatigue II. Use the AASHTO HL-93 loading, and assume an urban highway with 8,000 vehicles per lane per day.

Solution

For each limit state, the following limit state equation must be satisfied.

A Art. 1.3.2

$$\Sigma \eta_D \eta_R \eta_I \gamma_i Q_i \leq \Phi R_n = R_r$$

The following terms are defined for limit states.

A Eq. 1.3.2.1-1

Q = U = factored load effect ($U = \Sigma\, \eta_i \gamma_i Q_i \leq \Phi R_n = R_r$)

Q_i force effect from various loads

R_n nominal resistance

R_r factored resistance (ΦR_n)

γ_i load factor; a statistically based multiplier applied to force effects including distribution factors and load combination factors

η_i load modifier relating to ductility, redundancy, and operational importance 1.0 (for conventional designs)

A Art. 1.3.3

η_D ductility factor (strength only) 1.0 (for conventional designs)

A Art. 1.3.4

η_R redundancy factor (strength only) 1.0 (for conventional designs)

A Art. 1.3.5

η_I operational importance factor (strength and extreme only) 1.0 (for conventional designs)

Φ resistance factor, a statistically based multiplier applied to nominal resistance 1.0 (for Service Art. 1.3.2.1 and Fatigue Limit Art. C6.6.1.2.2 States)

Φ resistance factor For concrete structures see A Art. 5.5.4.2.1
For steel structures see A Art. 6.5.4.2

TABLE 2.8

Load Modifier Factors

Load Modifier	Strength	Service	Fatigue
Ductility, η_D	1.0	1.0	1.0
Redundancy, η_r	1.0	1.0	1.0
Operational importance, η_I	1.0	N/A	N/A
$\eta_i = \eta_D \eta_R \eta_I \geq 0.95$	1.0	1.0	1.0

Step 1: Select Load Modifiers, Combinations, and Factors

See Table 2.8.

A Arts. 1.3.3, 1.3.4, 1.3.5, 1.3.2.1; Eq. 1.3.2.1-2

Load Combinations and Factors

A Art. 3.4.1

$Q = \ddot{U}$ is the total factored force effect, $\Sigma\, \eta_i \gamma_i Q_i$.
Q_i represents the force effects from loads specified.
η represents specified load modifier.
γ_i represents specified load factors.

The following load combinations are used in this example.

A Tbls. 3.4.1-1, 3.4.1-2

Strength I Limit State: $Q = (1.25\ DC + 1.50\ DW + 1.75(LL + IM))$
Service I Limit State: $Q = (1.0(DC + DW) + 1.0(LL + IM))$
Fatigue II Limit State: $Q = (0.75)(LL + IM)$

Step 2: Determine Maximum Live Load Moments at Midspan

Please see Figures 2.45a–e.

A Art. 3.6.1.2

The HL-93 live load consists of the design lane load and either the design tandem or the design truck load (whichever is greater). The design truck is HS-20. The design tandem consists of a pair of 25.0 kip axles, spaced at 4.0 ft apart, and a transverse spacing of wheels is 6.0 ft. The design lane load consists of 0.64 kips/ft uniformly distributed in the longitudinal direction. Transversely, the design lane load is distributed uniformly over a 10.0 ft width, within a 12.0 ft design lane.

The force effects from the design lane load are not subject to a dynamic load allowance.

A Art. 3.6.2.1

The fatigue truck (HS-20) is placed in a single lane with a constant spacing of 30 ft between rear axles. See Figure 2.45e.

A Art. 3.6.1.4.1

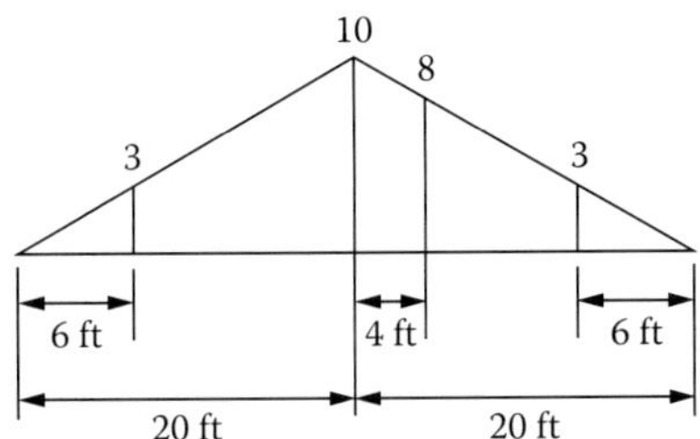

FIGURE 2.45a
Influence line for maximum moment at midspan.

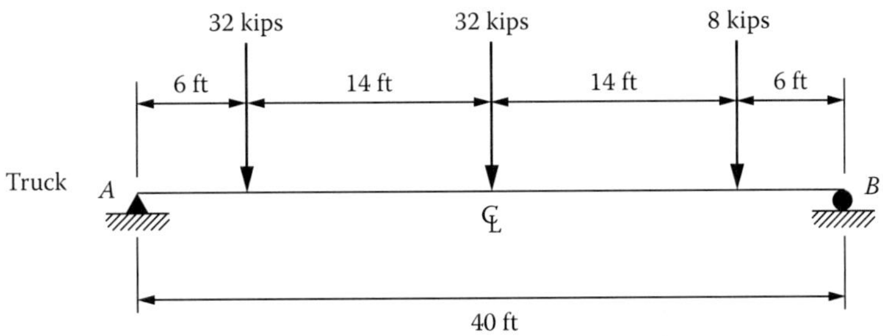

FIGURE 2.45b
Controlling load position for moment at midspan for design truck load (HS-20).

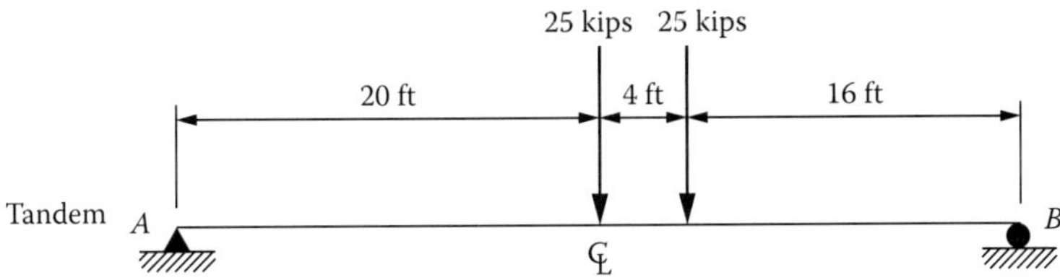

FIGURE 2.45c
Controlling load position for moment at midspan for design tandem load.

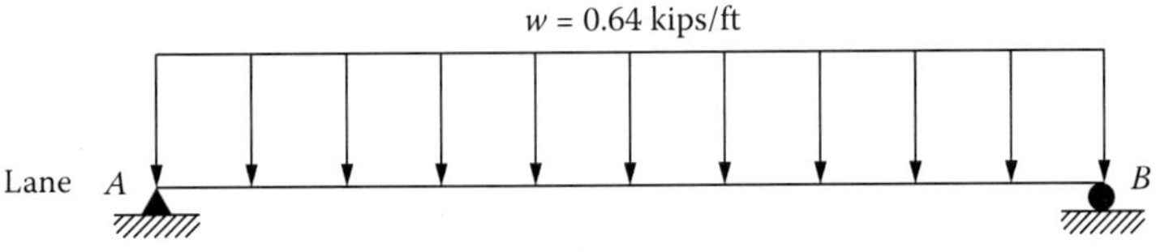

FIGURE 2.45d
Controlling load position for moment at midspan for design lane load.

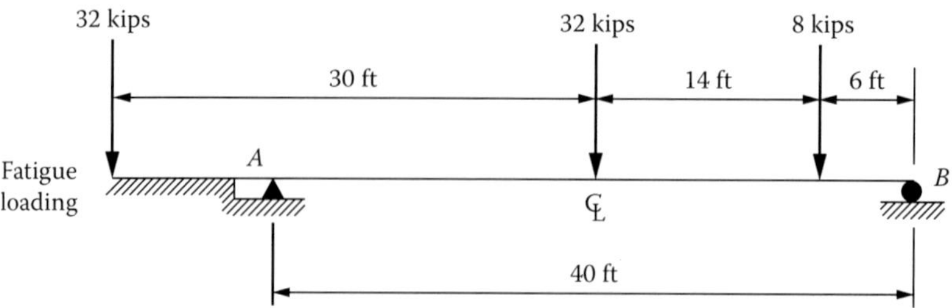

FIGURE 2.45e
Single lane fatigue load placement with one design truck load for maximum moment at mid-span. (**A Art. 3.6.1.4.1**)

Using the influence line diagrams (IL), Figure 2.45a

$$M_{tr} = (32\text{ kips})(10\text{ ft}) + (32\text{ kips} + 8\text{ kips})(3\text{ ft}) = 440\text{ ft-kips per lane}$$

$$M_{tandem} = (25\text{ kips})(10\text{ ft} + 8\text{ ft}) = 450\text{ ft-kips per lane [controls]}$$

$$M_{ln} = \frac{wL^2}{8} = \frac{\left(0.64\frac{\text{kips}}{\text{ft}}\right)(40\text{ ft})^2}{8} = 128\text{ ft-kips per lane}$$

$$M_{fatigue} = (32\text{ kips})(10\text{ ft}) + (8\text{ kips})(3\text{ ft}) = 344\text{ ft-kips per lane}$$

Step 3: Calculate Live Load Force Effects for Moment, Q_i

Where w is the clear roadway width between curbs and/or barriers, the number of design lanes is,

A Art. 3.6.1.1.1

$$N_L = \frac{w}{12} = \frac{28\text{ ft}}{12\frac{\text{ft}}{\text{lane}}} = 2.33\text{ lanes (2 lanes)}$$

The modulus of elasticity of concrete where $w_c = 0.145\text{ kips/in}^3$ is

A Eq. 5.4.2.4-1

$$E_c = (33{,}000)(w_c)^{1.5}\sqrt{f'_c}$$

$$= (33{,}000)\left(0.145\frac{\text{kips}}{\text{in}^3}\right)^{1.5}\sqrt{4\frac{\text{kips}}{\text{in}^2}}$$

$$= 3644\text{ kips/in}^2$$

The modular ratio between steel and concrete is

$$n = \frac{E_s}{E_c} = \frac{29{,}000 \dfrac{\text{kips}}{\text{in}^2}}{3644 \dfrac{\text{kips}}{\text{in}^2}}$$

$$= 7.96 \text{ (we will use 8)}$$

n modular ratio between steel and concrete 8

AISC Tbl. 1-1 for W 33 × 130

d	depth of beam	33.10 in
A	area of beam	38.3 in^2
I	moment of inertia of beam	6710 in^4
E_{beam}	modulus of elasticity of beam	29,000 kips/in^2
E_{deck}	modulus of elasticity of concrete deck	3644 kips/in^2
L	span length	40 ft
t_s	slab thickness	7 in
b_f	flange width	11.5 in
t_f	flange thickness	0.855 in
S_x	section modulus	406 in^3
t_w	web thickness	0.58 in

See Figure 2.46.

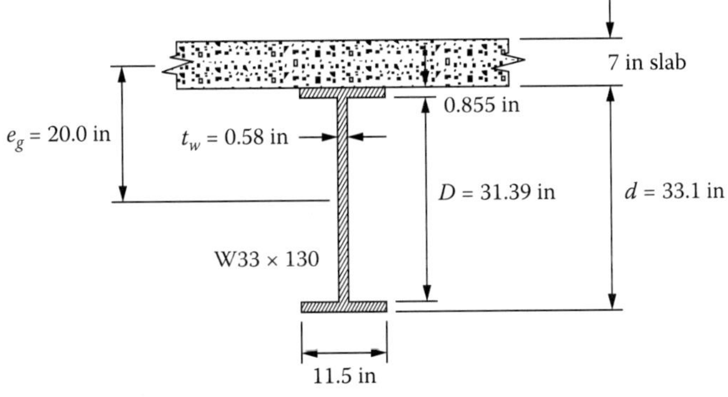

FIGURE 2.46
Noncomposite steel section at midspan.

The distance between the centers of the slab and steel beam is

$$e_g = \frac{d}{2} + \frac{t_s}{2} = \frac{33.1\text{ in}}{2} + \frac{7\text{ in}}{2} = 20.0\text{ in}$$

The longitudinal stiffness parameter, K_g, is

A Art. 4.6.2.2.1, A Eq. 4.6.2.2.1-1

$$K_g = n(I + Ae_g^2) = (8)(6710\text{ in}^4 + (38.3\text{ in}^2)(20\text{ in})^2) = 176{,}240\text{ in}^4$$

$$\frac{K_g}{12\,Lt_s^3} = \frac{176{,}240\text{ in}^4}{(12)(40\text{ ft})(7\text{ in})^3} \approx 1.0$$

Type of deck is (a).

A Tbl. 4.6.2.2.1-1

For interior beams with concrete decks, the live load moment may be determined by applying the lane fraction specified in AASHTO Table 4.6.2.2.2b-1, or Appendix A.

A Art. 4.6.2.2.2b

The multiple presence factor, m, applies to the lever rule case for live load distribution factors only.

A Art. 3.6.1.1.2

In the approximate equations for distribution factors, multiple factors are already included.

The distribution of live load moment per lane for interior beams of typical deck cross section (a) with one design lane loaded is

A Tbl. 4.6.2.2b-1 or Appendix A

$$DFM_{si} = \left(0.06 + \left(\frac{S}{14}\right)^{0.4}\left(\frac{S}{L}\right)^{0.3}\left(\frac{K_g}{12\,Lt_s^3}\right)^{0.1}\right)$$

$$= \left(0.06 + \left(\frac{8.33\text{ ft}}{14}\right)^{0.4}\left(\frac{8.33\text{ ft}}{40\text{ ft}}\right)^{0.3}(1.0)^{0.1}\right)$$

$$= 0.567\text{ lanes}$$

The distribution of live load moment per lane for interior beams with two or more design lanes loaded is

$$DFM_{mi} = \left(0.075 + \left(\frac{S}{9.5}\right)^{0.6}\left(\frac{S}{L}\right)^{0.2}\left(\frac{K_g}{12\,Lt_s^3}\right)^{0.1}\right)$$

$$= \left(0.075 + \left(\frac{8.33\text{ ft}}{9.5}\right)^{0.6}\left(\frac{8.33\text{ ft}}{40\text{ ft}}\right)^{0.2}(1.0)^{0.1}\right)$$

$$= 0.750 \text{ lanes [controls]}$$

For exterior beams with type (a) concrete decks, the live load moment may be determined by applying the lane fraction specified in Appendix B.

A Art. 4.6.2.2.2d; Tbl. 4.6.2.2.2d-1

Use the lever rule to find the distribution factor for moments for exterior beams with one design lane loaded. See Figure 2.47.

$$\Sigma M_{@b} = 0$$

$$R_a(100\text{ in}) = (0.5\text{ P})(22\text{ in}) + (0.5\text{ P})(94\text{ in})$$

$$R_a = 0.58\text{ P} = 0.58$$

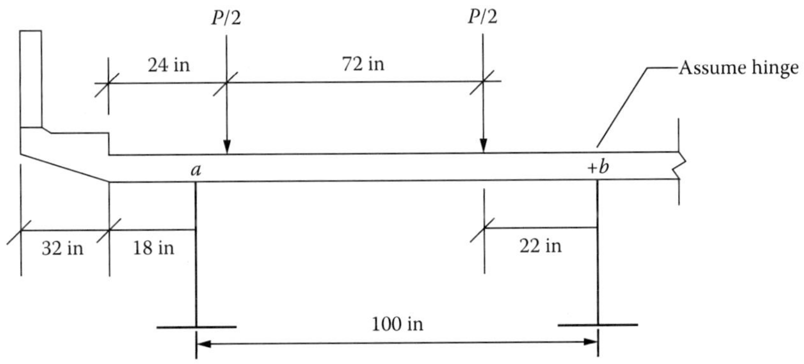

FIGURE 2.47
Lever rule for determination of distribution factor for moment in exterior beam, one lane loaded.

The distribution of the live load moment per lane for exterior beams of deck cross section type (a) with one design lane loaded is

A Tbl. 4.6.2.2.2d-1 or Appendix B

DFM_{se} = (multiple presence factor for one loaded lane) (R_a) = (1.2)(0.58) = 0.696

A Tbl. 3.6.1.1.2-1

Find the distribution factor for moments for exterior girders with two of more design lanes loaded, DFM_{me}, using g, an AASHTO distribution factor Table 4.6.2.2.2d-1 or Appendix B.

$$g = (e)(g_{interior})$$

$$g_{interior} = DFM_{mi} = 0.750$$

The distance from the exterior web of the exterior beam to the interior edge of the curb or traffic barrier, d_e, is 18 in (1.5 ft).

The correction factor for distribution is

$$e = 0.77 + \frac{d_e}{9.1} = 0.77 + \frac{1.5\text{ ft}}{9.1} = 0.935$$

The approximate load distribution factor for moments for exterior girders with two or more design lanes loaded has the multiple presence factors included.

A Art. 3.6.1.1.2

$$DFM_{me} = (e)(DFM_{mi}) = (0.935)(0.750) = 0.701$$

Step 4: Determine Maximum Live Load Shears

Please see Figures 2.48a–e.

Using the influence line diagram, Figure 2.48a

$$V_{tr} = (32\text{ kips})(1 + 0.65) + (8\text{ kips})(0.3) = 55.2\text{ kips [controls]}$$

$$V_{tandem} = (25\text{ kips})(1 + 0.9) = 47.5\text{ kips}$$

$$V_{ln} = \frac{wL}{2} = \frac{\left(0.64\frac{\text{kips}}{\text{ft}}\right)(40\text{ ft})}{2} = 12.8\text{ kips per lane}$$

$$V_{fatigue} = (32\text{ kips})(1 + 0.25) = 40.0\text{ kips}$$

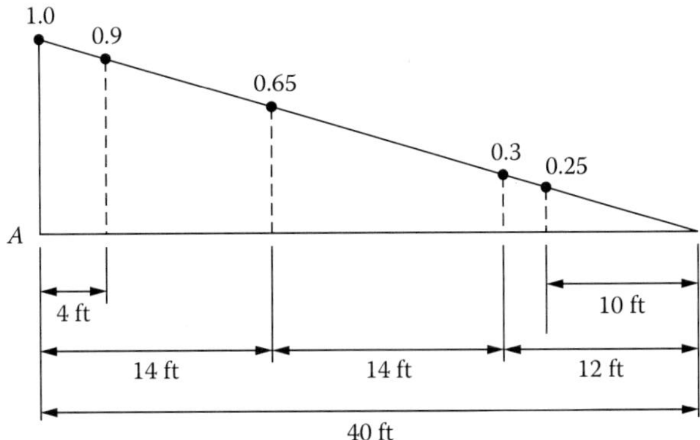

FIGURE 2.48a
Maximum live load shears; influence line for maximum shear at support.

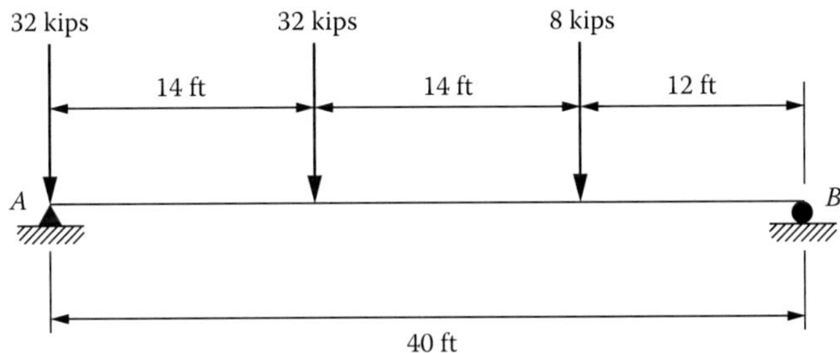

FIGURE 2.48b
Controlling load position for shear at support for design truck load (HS-20).

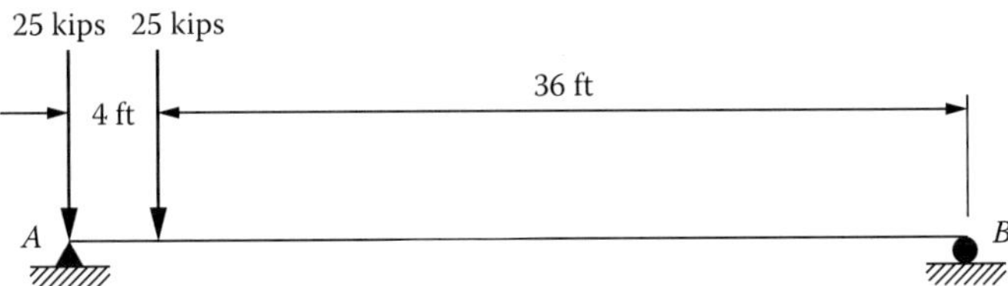

FIGURE 2.48c
Controlling load position for shear at support for design tandem load.

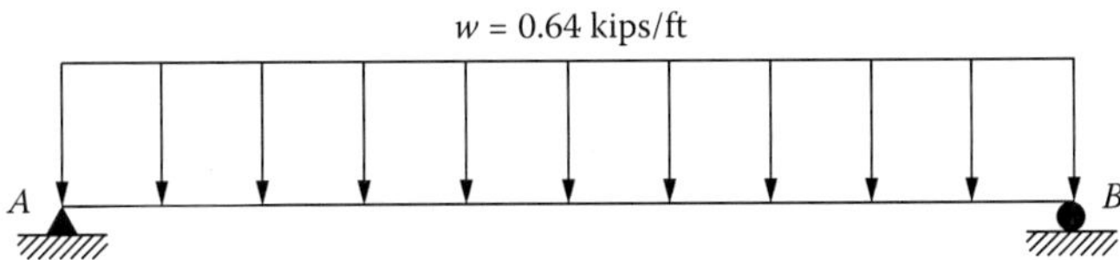

FIGURE 2.48d
Controlling load position for shear at support for design lane load.

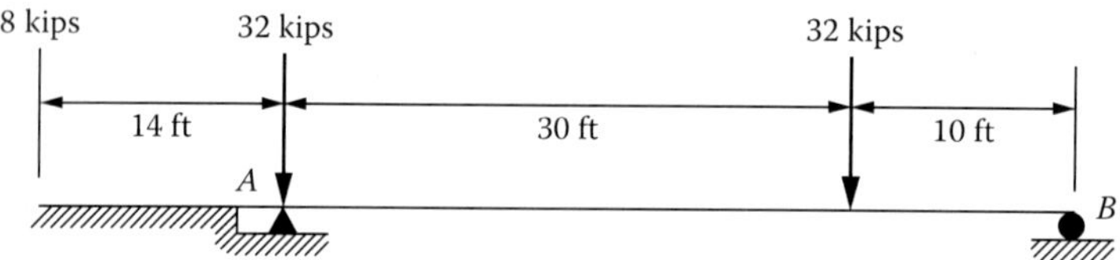

FIGURE 2.48e
Single lane fatigue load placement with one design truck load for maximum shear at support.

Step 5: Calculate the Live Load Force Effects for Shear, Q_i

A Arts. 4.6.2.2.3a and 3b; Tbls. 4.6.2.2.3a-1 and 3b-1

For interior beams with concrete decks, the live load shears may be determined by applying the lane fraction specified in Table 4.6.2.2.3a-1 or Appendix C.

The distribution for the live load shears for interior beams with one design lane loaded is

$$DFV_{si} = 0.36 + \left(\frac{S}{25}\right) = 0.36 + \left(\frac{8.33\text{ ft}}{25}\right) = 0.693\text{ lanes}$$

The distribution for the live load shears for interior beams with two or more design lanes loaded is

$$DFV_{mi} = 0.2 + \left(\frac{S}{12}\right) - \left(\frac{S}{35}\right)^2 = 0.2 + \left(\frac{8.33\text{ ft}}{12}\right) - \left(\frac{8.33\text{ ft}}{35}\right)^2 = 0.838\text{ lanes [controls]}$$

A Art. 4.6.2.2.3a; Tbl. 4.6.2.2.3a-1 or Appendix C

For exterior beams with type (a) concrete decks, the live load shears may be determined by applying the lane fraction in Table 4.6.2.2.3b-1 or Appendix D.

Using the lever rule, the distribution factor for live load shear for exterior beams with one design lane loaded is the same as the corresponding live load moment.

Therefore, $DFV_{se} = DFM_{se} = 0.696$ lanes.

Find the distribution factor for live load shear for exterior beams with two or more design lanes loaded. The correction factor for distribution is

$$e = 0.6 + \left(\frac{d_e}{10}\right) = 0.6 + \left(\frac{1.5\text{ ft}}{10}\right) = 0.75$$

$$DFV_{mi} = g = (e)(g_{interior}) = (e)(DFV_{mi}) = (0.75)(0.838) = 0.629\text{ lanes}$$

Please see Table 2.9.

TABLE 2.9

Summary of Distribution Factors:

	Load Case	DFM_{int}	DFM_{ext}	DFV_{int}	DFV_{ext}
Distribution Factors from A Art. 4.6.2.2.2	Multiple Lanes Loaded	0.750	0.701	0.838	0.629
	Single Lane Loaded	0.567	0.696	0.693	0.696
Design Value		0.750	0.701	0.838	0.696

TABLE 2.10

Summary of Fatigue Limit State Distribution Factors[a]

	Load Case	DFM_{int}	DFM_{ext}	DFV_{int}	DFV_{ext}
Distribution Factors from A Art. 4.6.2.2.2	Multiple Lanes Loaded	N/A	N/A	N/A	N/A
	Single Lane Loaded	0.567/1.2 = 0.473	0.696/1.2 = 0.580	0.693/1.2 = 0.578	0.696/1.2 = 0.580
Design Value		0.473	0.580	0.578	0.580

[a] Divide values above by multiple presence factor of 1.20.

When a bridge is analyzed for fatigue load using the approximate load distribution factors, one traffic lane shall be used. Therefore, the force effect shall be divided by 1.20.

Please see Table 2.10.

A Art. 3.6.1.1.2, 3.6.1.4.3b

Step 6: Calculate the Live Load Moments and Shears

The distributed live load moment per interior girder with multiple (two or more) lanes loaded governs is

A Art. 3.6.2.1

$$M_{LL+IM} = DFM_{int}((M_{tr} \text{ or } M_{tandem})(1 + IM) + M_{ln})$$

$$= (0.75)((450 \text{ ft-kips})(1 + 0.33) + 128 \text{ ft-kips})$$

$$= 545 \text{ ft-kips}$$

The distributed live load moment per exterior girder with multiple lanes loaded is

$$M_{LL+IM} = DFM_{ext}((M_{tr} \text{ or } M_{tandem})(1 + IM) + M_{ln})$$

$$= (0.701)((450 \text{ ft-kips})(1 + 0.33) + 128 \text{ ft-kips})$$

$$= 509.3 \text{ ft-kips}$$

A reduced dynamic load allowance of 15% is applied to the fatigue load.

A Tbl. 3.6.2.1-1

For a single lane loaded, the fatigue moment for interior beams is

$$M_{fatigue+IM} = \frac{DFM_{si}}{m} M_{fatigue}\left(LL + IM\right)$$

$$= \left(\frac{0.567}{1.2}\right)(344 \text{ ft-kips})(1.15)$$

$$= 187.0 \text{ ft-kips}$$

For a single lane loaded, the fatigue moment for exterior beams is

$$M_{fatigue+IM} = \frac{DFM_{se}}{m} M_{fatigue}\left(LL + IM\right)$$

$$= \left(\frac{0.696}{1.2}\right)(344 \text{ ft-kips})(1.15)$$

$$= 229.0 \text{ ft-kips per beam}$$

The distributed live load shear for interior beams is,

$$V_{LL+IM} = DFV((V_{tr} \text{ or } V_{tandem})(1 + IM) + V_{ln})$$

The distributed live load shear for interior beams ($DFV_{mi} = 0.838$) is

$$V_{LL+IM} = (0.838)((55.2 \text{ kips})(1 + 0.33) + 12.8 \text{ kips}) = 72.2 \text{ kips}$$

The distributed live load shear for exterior beams ($DFV_{se} = 0.696$) is

$$V_{LL+IM} = (0.696)((55.2 \text{ kips})(1 + 0.33) + 12.8 \text{ kips}) = 60.0 \text{ kips}$$

The fatigue shear for interior beams is

$$V_{fatigue+IM} = \frac{DFV_{si}}{m} V_{fatigue}\left(LL + IM\right)$$

$$= \left(\frac{0.693}{1.2}\right)(40 \text{ kips})(1.15)$$

$$= 26.6 \text{ kips}$$

The fatigue shear for exterior beams is,

$$V_{fatigue+IM} = \frac{DFV_{se}}{m} V_{fatigue}(LL + IM)$$

$$= \left(\frac{0.696}{1.2}\right)(40 \text{ kips})(1.15)$$

$$= 26.7 \text{ kips}$$

Step 7: Calculate Force Effects from the Dead Load and Wearing Surface

Find the force effects from the dead load and wearing surface for interior beams.

The nominal weight of the deck slab is

$$w_{slab,int} = \left(0.145 \frac{\text{kips}}{\text{ft}^3}\right)(7 \text{ in})\left(\frac{1 \text{ ft}}{12 \text{ in}}\right)(8.33 \text{ ft}) = 0.705 \text{ kips/ft}$$

The nominal weight of a W33 x 130 beam, w_{33x130}, is 0.130 kips/ft.

$$w_{DC} = 0.705 \text{ kips/ft} + 0.130 \text{ kip/ft} = 0.835 \text{ kips/ft}$$

The nominal weight of the wearing surface is,

$$w_{DW} = \left(0.140 \frac{\text{kips}}{\text{ft}^3}\right)(3 \text{ in})\left(\frac{1 \text{ ft}}{12 \text{ in}}\right)(8.33 \text{ ft}) = 0.292 \text{ kips/ft}$$

The interior beam moments and shears are,

$$V_{DC} = \frac{w_{DC}L}{2} = \frac{\left(0.835 \frac{\text{kips}}{\text{ft}}\right)(40 \text{ ft})}{2} = 16.7 \text{ kips}$$

$$M_{DC} = \frac{w_{DC}L^2}{8} = \frac{\left(0.835 \frac{\text{kips}}{\text{ft}}\right)(40 \text{ ft})^2}{8} = 167 \text{ ft-kips}$$

TABLE 2.11

Summary of Loads, Shears, and Moments in Interior Beams

Load Type	w (kip/ft)	Moment (ft-kips)	Shear (kips)
DC	0.835	167	16.7
DW	0.292	58.4	5.84
LL + IM	N/A	545	72.2
Fatigue + IM	N/A	187	26.6

$$V_{DW} = \frac{w_{DW}L}{2} = \frac{\left(0.292\,\frac{\text{kips}}{\text{ft}}\right)(40\text{ ft})}{2} = 5.84\text{ kips}$$

$$M_{DW} = \frac{w_{DW}L^2}{8} = \frac{\left(0.292\,\frac{\text{kips}}{\text{ft}}\right)(40\text{ ft})^2}{8} = 58.4\text{ ft-kips}$$

Table 2.11 summarizes the loads, shears, and moments for interior beams.

Find the force effects from the dead load and wearing surface for exterior beams.

The nominal weight of the deck slab is

$$w_{slab,ext} = \left(0.145\,\frac{\text{kips}}{\text{ft}^3}\right)(7\text{ in})\left(\frac{1\text{ ft}}{12\text{ in}}\right)\left(\frac{8.33\text{ ft}}{2} + 1.5\text{ ft}\right) = 0.479\text{ kips/ft}$$

The nominal weight of a W33 x 130 beam, $w_{W33x130}$, is 0.130 kips/ft. Assume that the nominal weight of the barrier, sidewalk, and railings, $w_{barrier+sidewalk+rail}$, is 1.0 kips/ft. For exterior girders, the distributed load for the slab, beam, barrier, sidewalk, and girders is

$$w_{DC} = w_{slab,ext} + w_{W33x130} + w_{barrier+sidewalk+rail}$$

$$= 0.479\text{ kips/ft} + 0.130\text{ kips/ft} + 1.0\text{ kips/ft} = 1.61\text{ kips/ft}$$

The nominal weight of the wearing surface is

$$w_{DW} = \left(0.140\,\frac{\text{kips}}{\text{ft}^3}\right)(3\text{ in})\left(\frac{1\text{ ft}}{12\text{ in}}\right)\left(\frac{8.33\text{ ft}}{2} + 1.5\text{ ft}\right) = 0.198\text{ kips/ft}$$

The exterior beam moments and shears are

$$V_{DC} = \frac{w_{DC}L}{2} = \frac{\left(1.61\,\frac{\text{kips}}{\text{ft}}\right)(40\text{ ft})}{2} = 32.2\text{ kips}$$

TABLE 2.12

Summary of Loads, Shears, and Moments in Exterior Beams

Load Type	w (kip/ft)	Moment (ft-kips)	Shear (kips)
DC	1.61	322	32.2
DW	0.198	39.6	3.96
LL + IM	N/A	509.3	60.0
Fatigue + IM	N/A	229	26.7

$$M_{DC} = \frac{w_{DC}L^2}{8} = \frac{\left(1.61\frac{\text{kips}}{\text{ft}}\right)(40\text{ ft})^2}{8} = 322\text{ ft-kips}$$

$$V_{DW} = \frac{w_{DW}L}{2} = \frac{\left(0.198\frac{\text{kips}}{\text{ft}}\right)(40\text{ ft})}{2} = 3.96\text{ kips}$$

$$M_{DW} = \frac{w_{DW}L^2}{8} = \frac{\left(0.198\frac{\text{kips}}{\text{ft}}\right)(40\text{ ft})^2}{8} = 39.6\text{ ft-kips}$$

Table 2.12 summarizes the loads, shears, and moments for exterior beams.

Step 8: Find the Factored Moments and Shears with Applicable Limit States

A Tbls. 3.4.1-1, 3.4.1-2

Strength I Limit State for Interior Beam

$$U = (1.25\text{ DC} + 1.50\text{ DW} + 1.75(\text{LL} + \text{IM}))$$

$$V_u = ((1.25)(16.7\text{ kips}) + (1.50)(5.84\text{ kips}) + (1.75)(72.2\text{ kips}))$$

$$= 156\text{ kips}$$

$$M_u = ((1.25)(167\text{ ft-kips}) + (1.50)(58.4\text{ ft-kips}) + (1.75)(545\text{ kips}))$$

$$= 1250\text{ ft-kips}$$

Service I Limit State

$$U = (1.0(\text{DC} + \text{DW}) + 1.0(\text{LL} + \text{IM}))$$

$V_u = ((1.0)(16.7 \text{ kips} + 5.84 \text{ kips}) + (1.00)(72.2 \text{ kips}))$

$= 94.74 \text{ kips}$

$M_u = ((1.0)(167 \text{ ft-kips} + 58.4 \text{ ft-kips}) + (1.00)(545 \text{ kips}))$

$= 770.4 \text{ ft-kips}$

Fatigue II Limit State for Interior Beams (with dead loads considered to be conservative)

$U = \eta(1.0\ DC + 1.0\ DW + 0.75(LL + IM))$

$V_u = (1.0)((1.0)(16.7 \text{ kips}) + (1.0)(5.84 \text{ kips}) + (0.75)(26.6 \text{ kips}))$

$= 42.5 \text{ kips}$

$M_u = (1.0)((1.0)(167 \text{ ft-kips}) + (1.0)(58.4 \text{ ft-kips}) + (0.75)(187 \text{ kips}))$

$= 366 \text{ ft-kips}$

Strength I Limit State for Exterior Beam

A Tbls. 3.4.1-1, 3.4.1-2

$U = (1.25\ DC + 1.50\ DW + 1.75(LL + IM))$

$V_u = ((1.25)(32.2 \text{ kips}) + (1.50)(3.96 \text{ kips}) + (1.75)(60.0 \text{ kips}))$

$= 151.0 \text{ kips}$

$M_u = ((1.25)(322 \text{ ft-kips}) + (1.50)(39.6 \text{ ft-kips}) + (1.75)(509.3 \text{ kips}))$

$= 1353.2 \text{ ft-kips}$ [controls]

Service I Limit State

$U = (1.0(DC + DW) + 1.0(LL + IM))$

$V_u = ((1.0)(32.2 \text{ kips} + 3.96 \text{ kips}) + (1.00)(60.0 \text{ kips}))$

$= 96.2 \text{ kips}$

$M_u = ((1.0)(322 \text{ ft-kips} + 39.6 \text{ ft-kips}) + (1.00)(509.3 \text{ kips}))$

$= 870.9 \text{ ft-kips}$

Fatigue II Limit State for Exterior Beams (with dead load considered to be conservative)

$$U = (1.0\ DC + 1.0\ DW + 0.75(LL + IM))$$

$$V_u = ((1.0)(32.2\text{ kips}) + (1.0)(3.96\text{ kips}) + (0.75)(26.7\text{ kips}))$$

$$= 56.2\text{ kips}$$

$$M_u = ((1.0)(322\text{ ft-kips}) + (1.0)(39.6\text{ ft-kips}) + (0.75)(229\text{ kips}))$$

$$= 533.4\text{ ft-kips}$$

Step 9: Check Fundamental Section Properties for Strength Limit I

For the Strength I Limit State, the exterior beam moment controls (please refer to Step 7):

The maximum factored moment, M_u, is 1353.2 ft-kips.

For noncomposite sections,

A Art. C6.10.1.2; A App. A6.1.1, A6.1.2; Art. 6.5.4.2

$$M_u \le \Phi_f M_n$$

Because M_n is the required nominal moment resistance and Φ_f is 1.0 for steel structures, M_n must be at least M_u = 1353.2 ft-kips.

Check the plastic section modulus required.

$$Z_{req'd} \ge \frac{M_u}{F_y} = \frac{(1353.2\text{ ft-kips})\left(12\frac{\text{in}}{\text{ft}}\right)}{50\frac{\text{kips}}{\text{in}^2}} = 324.7\text{ in}^3$$

$$Z_{W33x130} = 467\text{ in}^3 \ge 324.7\text{ in}^3\ [\text{OK}]$$

AISC Tbl.1-1

Check the web proportions.

A Art. 6.10.2.1.1, A Eq. 6.10.2.1.1-1

$$\frac{D}{t_w} = \frac{33.1\ \text{in} - (2)(0.855\ \text{in})}{0.580\ \text{in}} = 54.1 \le 150\ \ [\text{OK}]$$

Check the flange proportions.

A Art. 6.10.2.2, A Eq. 6.10.2.2-1; A Eq. 6.10.2.2-2

$$\frac{b_f}{2t_f} = \frac{11.5\ \text{in}}{(2)(0.855\ \text{in})} = 6.73 \le 12.0\ \ [\text{OK}]$$

$$b_f \ge \frac{D}{6}$$

$$\frac{D}{6} = \frac{33.1\ \text{in} - (2)(0.855\ \text{in})}{6}$$

$$= 5.23\ \text{in} \le b_f = 11.5\ \text{in}\ [\text{OK}]$$

$$t_f \ge 1.1\ t_w$$

A Eq. 6.10.2.2-3

$$1.1\ t_w = (1.1)(0.580\ \text{in}) = 0.638$$

$$t_f = 0.855\ \text{in} \ge 0.638\ [\text{OK}]$$

Compare the moment of inertia of the compression flange of the steel section about the vertical axis in the plane of the web, I_{yc}, to the moment of inertia of the tension flange of the steel section about the vertical axis in the plane of the web.

$$0.1 \le \frac{I_{yc}}{I_{yt}} \le 10$$

A Eq. 6.10.2.2-4

$$0.1 \le 1.0 \le 10\ [\text{OK}]$$

Check the material thickness.

A Art. 6.7.3

$$t_w = 0.580\ \text{in} \ge 0.25\ \text{in}\ [\text{OK}]$$

Step 10: Check Live Load Deflection

A Art. 2.5.2.6.2; A Art. 3.6.1.3.2

The live load deflection should be taken as the larger of:

- That resulting from the design truck alone, or
- That resulting from 25% of the design truck plus the design lane load

With Service I load combination, the maximum live load deflection limit is span/800.

The allowable service load deflection, which must be less than or equal to the span/800, is

A Art. 2.5.2.6.2

$$\frac{(40\ \text{ft})\left(12\,\frac{\text{in}}{\text{ft}}\right)}{800} = 0.6\ \text{in}$$

For a straight multibeam bridge, the distribution factor for deflection, $DF_{deflection}$, is equal to the number of lanes divided by the number of beams.

A Comm. 2.5.2.6.2

$$DF_{deflection} = \left(\frac{\text{no. of lanes}}{\text{no. of beams}}\right) = \left(\frac{2}{4}\right) = 0.5$$

Deflection is governed by either design truck loading (HS-20) alone, or 25% of truck plus design lane load. See Figure 2.49.

A Art. 3.6.1.3.2

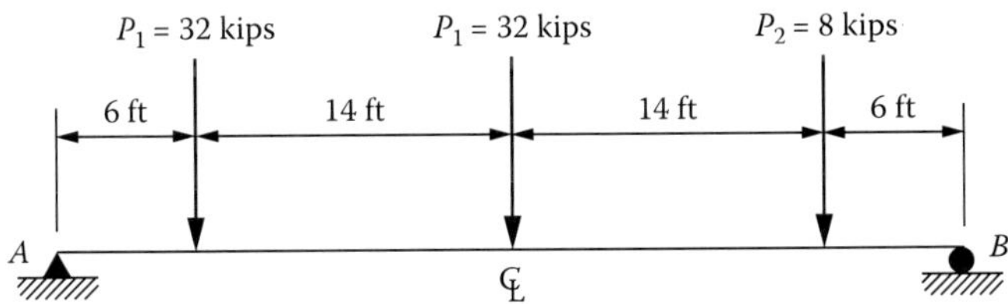

FIGURE 2.49
Position of design truck loading (HS-20) for deflection at midspan.

$$P_{1,beam} = \left(DF_{deflection}\right)\left(P_1\right)\left(1 + \frac{IM}{100}\right) = (0.5)(32\ \text{kips})(1 + 0.33) = 21.3\ \text{kips}$$

$$P_{2,beam} = \left(DF_{deflection}\right)\left(P_2\right)\left(1 + \frac{IM}{100}\right) = (0.5)(8\ \text{kips})(1 + 0.33) = 5.32\ \text{kips}$$

$$\Delta_{truck} = \Delta P_{1,beam} + \Delta P_{2,beam} + \Delta P_{1,beam}$$

$$\Delta_{truck} = \Sigma\left(\frac{Pbx}{eEIL}\left(L^2 - b^2 - x^2\right)\right) + \frac{P_1 L^3}{48\ EI}$$

AISC Tbl. 3-23

$$= \left(\frac{(21.3\ \text{kips})(240\ \text{in})(72\ \text{in})}{(6)\left(29{,}000\ \frac{\text{kips}}{\text{in}^2}\right)(6710\ \text{in}^4)(480\ \text{in})}\left((480\ \text{in})^2 - (408\ \text{in})^2 - (72\ \text{in})^2\right)\right)$$

$$+ \left(\frac{(5.32\ \text{kips})(72\ \text{in})(240\ \text{in})}{(6)\left(29{,}000\ \frac{\text{kips}}{\text{in}^2}\right)(6710\ \text{in}^4)(480\ \text{in})}\left((480\ \text{in})^2 - (72\ \text{in})^2 - (408\ \text{in})^2\right)\right)$$

$$+ \left(\frac{(21.3\ \text{kips})(480\ \text{in})^3}{48\left(29{,}000\ \frac{\text{kips}}{\text{in}^2}\right)(6710\ \text{in}^4)}\right)$$

$= 0.039\ \text{in} + 0.010\ \text{in} + 0.252\ \text{in} = 0.301\ \text{in}$ [controls]

Check requirements for 25% of truck and design lane load (0.64 kip/ft uniformly distributed).

$$\Delta_{lane} = \frac{5\ wL^4}{384\ EI} = \frac{(5)\left(0.64\ \frac{\text{kips}}{\text{ft}}\left(\frac{1\ \text{ft}}{12\ \text{in}}\right)\right)(480\ \text{in})^4}{(384)\left(29{,}000\ \frac{\text{kips}}{\text{in}^2}\right)(6710\ \text{in}^4)} = 0.189\ \text{in}$$

$$0.25\Delta_{truck} + \Delta_{lane} = (0.25)(0.301\ \text{in}) + 0.189\ \text{in} = 0.264\ \text{in}$$

The controlling deflection is

$$\Delta_{truck} = 0.301\ \text{in} \le 0.6\ \text{in}\ [\text{OK}]$$

Step 11: Check the Service Limit State

Permanent Deformation

A Art. 6.10.4.2

Service II load combination should apply to calculate stresses in structural steel section alone.

A Art. 6.10.4.2.1

Flanges shall satisfy the requirements for steel flanges of noncomposite sections.

A Art. 6.10.4.2.2; Eq. 6.10.4.2.2-3

$$f_f + \frac{f_l}{2} \le 0.80\ R_h F_{yf}$$

where:

f_f = flange stress due to Service II loads

R_h = hybrid factor = 1.0

A Art. 6.10.1.10.1

F_{yf} = minimum yield strength of flange

f_l = flange lateral bending stress ≅ 0.0

The flange lateral bending stress, f_l, is 0 because lateral bending stresses are assumed small.

$$\begin{aligned} 0.80\ R_h F_{yf} &= (0.80)(1.0)(50\text{ ksi}) \\ &= 40\text{ ksi} \end{aligned}$$

The factored Service II moment for exterior beams is

$$\begin{aligned} M_u &= 1.0(DC + DW) + 1.30(LL + IM) \\ &= 1.0(322\text{ ft-kips} + 39.6\text{ ft-kips}) + 1.30(509.3\text{ ft-kips}) \\ &= 1023.7\text{ ft-kips (controls)} \end{aligned}$$

The factored Service II moment for interior beams is

$$\begin{aligned} M_u &= 1.0\ ({}^{16}\eta\text{ ft-kips} + 58.4\text{ ft-kips}) + 1.30\ (545\text{ ft-kips}) \\ &= 933.8\text{ ft-kips} \end{aligned}$$

The section modulus, S, for a W33 × 130 is 406 in^3.

AISC Tbl.1-1

The flange stress due to the Service II loads calculated without consideration for flange bending is

$$\begin{aligned} f_f = \frac{M}{S} &= \frac{(1023.7\text{ ft-kips})\left(\dfrac{12\text{ in}}{1\text{ ft}}\right)}{406\text{ in}^3} \\ &= 30.25\text{ kips/in}^2 \le 0.80\ R_h F_{yf} = 40\text{ ksi [OK]} \end{aligned}$$

Step 12: Check Fatigue and Fracture Limit State

A Art. 6.6.1.2.2; A Art. 6.10.5.1

For load-induced fatigue considerations, each detail shall satisfy:

A Eq. 6.6.1.2.2-1

$$\gamma(\Delta f) \le (\Delta F)_n$$

where:

γ = load factor for the fatigue load combination
= 0.75 for Fatigue II Limit State

A Tbl. 3.4.1-1

(Δf) = force effect, live load stress range due to the fatigue load

A Art. 3.6.1.4

$(\Delta F)_n$ = nominal fatigue resistance (ksi)

A Art. 6.6.1.2.5

(ΔF_{TH}) = constant amplitude fatigue threshold (ksi)

A Tbl. 6.6.1.2.5-3

For the Fatigue II load combination and finite life,

A Eq. 6.6.1.2.5-2

$$(\Delta F)_n = \left(\frac{A}{N}\right)^{1/3}$$

The constant taken for Detail Category A, A_{out}, is 250 × 10^8 kips/in^2.

A Tbl. 6.6.1.2.5-1

The number of stress range cycles per truck passage, n, is 2.0 for simple span girders with span less than or equal to 40 ft.

A Tbl. 6.6.1.2.5-2

The number of design lanes, N_L, is 2.0.

The constant amplitude threshold $(\Delta F)_{TH}$, for Detail Category A, is 24.0 kips/in^2.

A Tbl. 6.6.1.2.5-3

$$\frac{1}{2}(\Delta F)_{TH} = \left(\frac{1}{2}\right)\left(24.0\frac{\text{kips}}{\text{in}^2}\right) = 12\text{kips/in}^2$$

The percentage of trucks (compared to all vehicles) in traffic for an urban interstate is 15%.

A Tbl. C3.6.1.4.2-1

This design example assumes that the average daily traffic, ADT, is 8,000 vehicles per day.

The average daily truck traffic is

$$ADTT = (0.15)(ADT) = (0.15)(8000 \text{ vehicles})(2 \text{ lanes}) = 2400 \text{ trucks per day}$$

The fraction of truck traffic in a single lane, p, is 0.85 when there are two lanes available to trucks.

A Tbl. 3.6.1.4.2-1

The single-lane average daily truck traffic, $ADTT_{SL}$, is

$$ADTT_{SL} = (p)ADTT = (0.85)(2400 \text{ trucks per day})$$

A Eq. 3.6.1.4.2-1

$$= 2040 \text{ trucks per lane per day}$$

The number of cycles of stress range for 75 years, N, is

A Eq. 6.6.1.2.5-3

$$N = (365 \text{ days})(75 \text{ years})(n)AADT_{SL}$$

where n = number of stress range cycles per truck passage = 2.0 for span lengths ≧ 40 ft

A Tbl. 6.6.1.2.5-2

$$N = (365 \text{ days})(75 \text{ years})(2.0 \text{ cycles per pass})(2040 \text{ trucks per day})$$

$$= 1.12 \times 10^8 \text{ cycles}$$

The nominal fatigue resistance, $(\Delta F)_n$, is

A Eq. 6.6.1.2.5-2

$$(\Delta F)_n = \left(\frac{A}{N}\right)^{1/3} = \left(\frac{250\text{x}10^8 \frac{\text{kips}}{\text{in}^2}}{1.12\text{x}10^8 \frac{\text{kips}}{\text{in}^2}}\right) = 6.07 \text{ kips/in}^2$$

Apply the fatigue limit state (excluding dead loads) using the maximum fatigue moment (exterior beam controls).

A Tbl. 3.4.1-1

$$Q = \gamma M$$
$$Q = \gamma[LL + IM]$$
$$Q = ((0.75)(LL + IM)) = ((0.75)(229 \text{ ft-kips})) = 172 \text{ ft-kips}$$

$$\gamma(\Delta f) = \frac{\gamma M}{S} = \frac{(172 \text{ ft-kips})\left(\dfrac{12 \text{ in}}{1 \text{ ft}}\right)}{406 \text{ in}^3}$$

$\gamma(\Delta f) = 5.08 \text{ kips/in}^2 \leq (\Delta F)_n = 6.07 \text{ kips/in}^2$ [OK]

A Eq. 6.6.1.2.2-1

Special fatigue requirement for the web.

A Art. 6.10.5.3

$$V_u \leq V_{cr}$$

A Eq. 6.10.5.3-1

where:
V_u = shear in the web due to the unfactored permanent load plus the factored fatigue load
V_{cr} = shear-buckling resistance

A Eq. 6.10.9.3.3-1

Exterior girder governs in fatigue shear, thus use exterior girder distribution factor and shears.

The shear in the web at the section under consideration due to the unfactored permanent load plus the factored fatigue load is

$$V_u = V_{DC} + V_{DW} + V_{fat}$$

$$V_{fat} = \gamma Q$$

$$= (0.75)(V_{fatigue+IM})$$

$$= (0.75)(26.7 \text{ kips})$$

$$= 20.0 \text{ kips}$$

$$V_u = 32.2. \text{ kips} + 3.96 \text{ kips} + 20.0 \text{ kips}$$

$$= 56.16 \text{ kips}$$

Nominal resistance of unstiffened webs is

A Art. 6.10.9.2, 6.10.9.3.3

$$V_n = V_{cr} = CV_p$$

A Eq. 6.10.9.3.3-1

where:
$V_p = 0.58\ F_{yw}Dt_w$
$= (0.58)(50\ \text{ksi})(31.39\ \text{in})(0.58\ \text{in})$
$= 528\ \text{kips}$
C = ratio of the shear-buckling resistance to the shear yield strength.

A Art. 6.10.9.3.2

Find the ratio of shear buckling resistance to the shear yield strength, C, by satisfying the following equation:

$$\text{If } \frac{D}{t_w} \le 1.12\sqrt{\frac{Ek}{F_{yw}}},\ C = 1.0$$

A Eq. 6.10.9.3.2-4

The shear-buckling coefficient k is

A Eq. 6.10.9.3.2-7

$$k = 5 + \frac{5}{\left(\frac{d_o}{D}\right)^2}$$

There are no specified transverse stiffeners, therefore d_o = infinity and k = 5.

$$\frac{D}{t_w} = \frac{33.1\ \text{in} - (2)(0.855\ \text{in})}{0.580\ \text{in}} = 54.1$$

$$1.12\sqrt{\frac{Ek}{F_{yw}}} = 1.12\sqrt{\frac{\left(29{,}000\frac{\text{kips}}{\text{in}^2}\right)(5)}{50\frac{\text{kips}}{\text{in}^2}}} = 60.3 > \frac{D}{t_w} = 54.1\ \ [\text{OK}]$$

Therefore, C = 1.0.

The nominal shear resistance, V_n, of the web is

A Eq. 6.10.9.3.3-1

$$V_n = V_{cr} = CV_p = (1.0)(528\ \text{kips})$$
$$= 528\ \text{kips}$$
$$V_u = 56.16\ \text{kips} \le V_n = V_{cr} = 528\ \text{kips}\ [\text{OK}]$$

A Eq. 6.10.5.3-1

Step 13: Check the Shear Adequacy at Support

Check that at the strength limit state the following requirement for shear is satisfied

$$V_u \le \Phi_v V_n$$

A Eq. 6.10.9.1-1

The nominal shear resistance, V_n, is 528 kips.

Φ_v is 1.0 for steel structures.

For the Strength I Limit State, V_u, was determined in Step 8.

$$V_u = 156 \text{ kips [interior beam controls]}$$

$$V_u = 156 \text{ kips} \le \Phi_v V_n = (1.0)(528 \text{ kips})$$

Design Example 5: Reinforced Concrete Slabs

Situation

The cast-in-place concrete deck for a simple span composite bridge is continuous across five steel girders, as shown in Figure 2.50.

The overall width of the bridge is 48 ft.

The clear roadway width is 44 ft, 6 in.

The roadway is a concrete slab 9 in thick, with a concrete compressive strength at 28 days, f'_c, of 4.5 kips/in^2, and a specified minimum yield strength of the steel, F_y, of 60 kips/in^2.

The steel girders are spaced at 10 ft as shown.

Allow for a 3 in future wearing surface, FWS, at 0.03 kips/ft^2.

Assume the area of the curb and parapet on each side is 3.37 ft^2.

The concrete self-weight, w_c, is equal to 150 lbf/ft^3.

Requirements

Design and review the reinforced concrete slab by the approximate method of analysis, AASHTO Art. 4.6.2.1. Use HL-93 loading.

A Arts. 4.6.2, 4.6.2.1

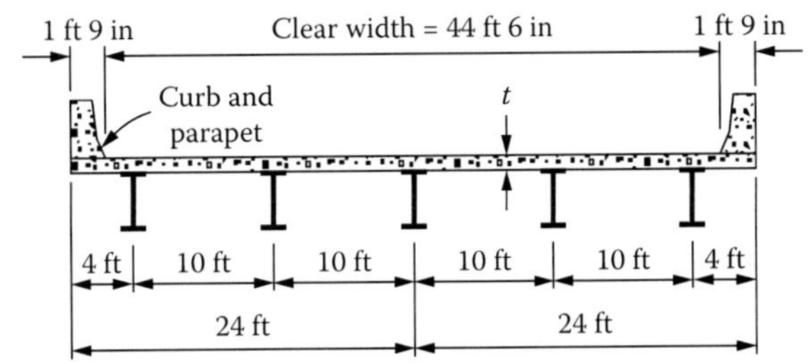

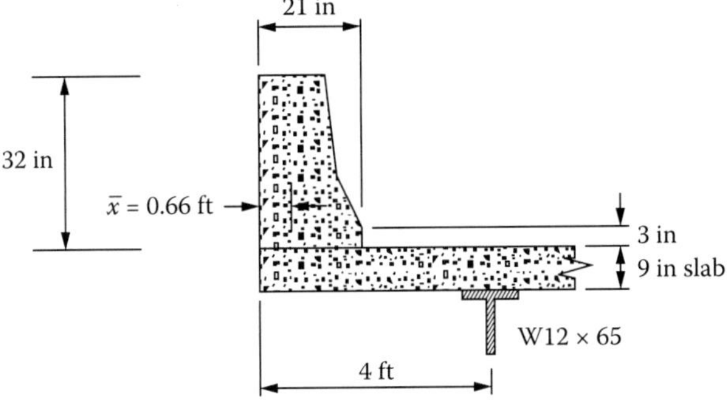

FIGURE 2.50
Concrete deck slab design example.

Solution

Step 1: Determine the Minimum Thickness of the Slab

The depth of a concrete deck should not be less than 7.0 in.

A Art. 9.7.1.1

The assumed slab thickness is t = 8.5 in + 0.5 in for integral wearing surface. Use t = 9 in.

Step 2: Determine Dead Loads, w

The following dead loads are determined for a 1.0 ft wide transverse strip.

For a 9.0 in thick slab, including cantilever, the dead load is

$$w_{slab} = \left(0.150\frac{\text{kips}}{\text{ft}^3}\right)(9.0\text{ in})\left(\frac{1\text{ ft}}{12\text{ in}}\right) = 0.113\text{ kips/ft}^2$$

The dead load for the future wearing surface, w_{FWS}, is given as 0.03 kips/ft^2. The dead load for the curb and parapet in each side is

$$w_{C\&P} = \left(0.150\frac{\text{kips}}{\text{ft}^3}\right)\left(3.37\ \text{ft}^2\right) = 0.506\ \text{kips/ft}\ \text{ of bridge span}$$

Step 3: Find the Dead Load Force Effects

Approximate elastic method analysis as specified in Art. 4.6.2.1 is permitted.

A Art. 9.6.1

The extreme positive moment in any deck panel between girders shall be taken to apply to all positive moment regions. Similarly, the extreme negative moment over any girder shall be taken to apply to all negative moment regions.

A Art. 4.6.2.1.1

The strips shall be treated as continuous beams with span lengths equal to the center-to-center distance between girders. For the purpose of determining force effects in the strip, the supporting components (i.e., girders) shall be assumed to be infinitely rigid. The strips should be analyzed by classical beam theory.

A Art. 4.6.2.1.6

Apply loadings and determine the reaction at point A, as well as moments at points A, B, and C. These values will be representative of the maximum reaction, as well as the maximum positive and negative moments for the continuous beam. Point A is located at the exterior support, point B at 0.4S inside the exterior support (4 ft to the right of point A), and point C at the first interior support. It is noted that in the continuous beam, four equal spans and all spans loaded, the maximum positive moment in the first interior span occurs at 0.40S (*AISC*, 13th edition, continuous beam, four equal spans, all spans loaded). For clarification, these points are approximately located as shown in Figure 2.51.

For this analysis, the resultant reaction and moments can be calculated using the moment distribution method, influence line design aids, or a computer software program. In this example, a structural engineering software

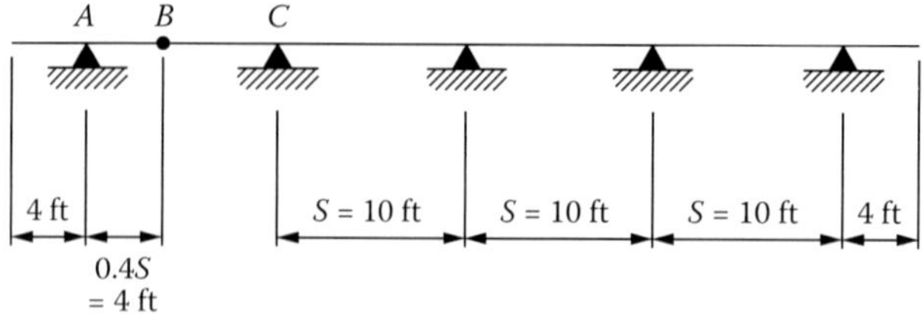

FIGURE 2.51
Locations in slab strips for maximum reactions and moments due to dead loads.

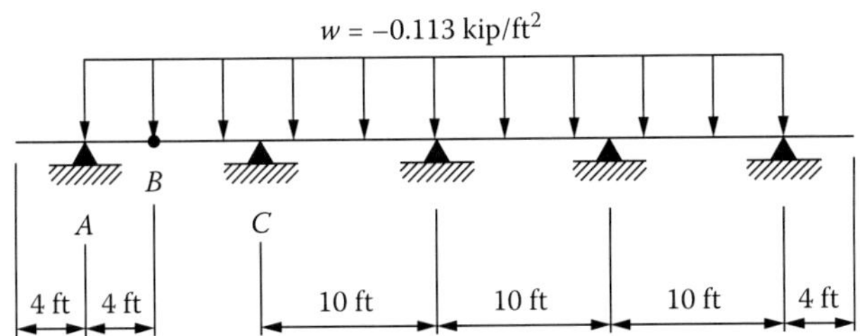

FIGURE 2.52
Moments and reactions for deck slab dead load excluding deck cantilever.

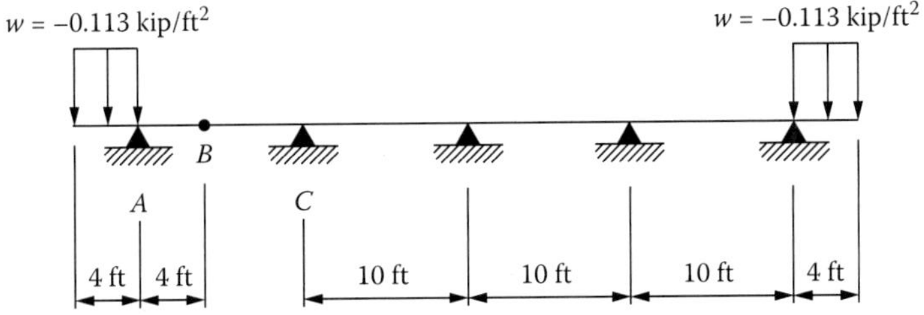

FIGURE 2.53
Moments and reaction for deck slab dead load in deck cantilever.

program was used to model and analyze the continuous slab. The loadings and results are summarized below.

The deck slab moments and reaction (excluding cantilever) are as follows and as shown in Figure 2.52.

$$M_A = 0 \text{ kip-ft/ft}$$

$$M_B = +0.872 \text{ kip-ft/ft}$$

$$M_C = -1.211 \text{ kip-ft/ft}$$

$$R_A = 0.444 \text{ kips}$$

The cantilever slab moments and reaction are as follows and as shown in Figure 2.53.

$$M_A = -0.904 \text{ kip-ft/ft}$$

$$M_B = -0.439 \text{ kip-ft/ft}$$

$$M_C = +0.258 \text{ kip-ft/ft}$$

$$R_A = 0.568 \text{ kips}$$

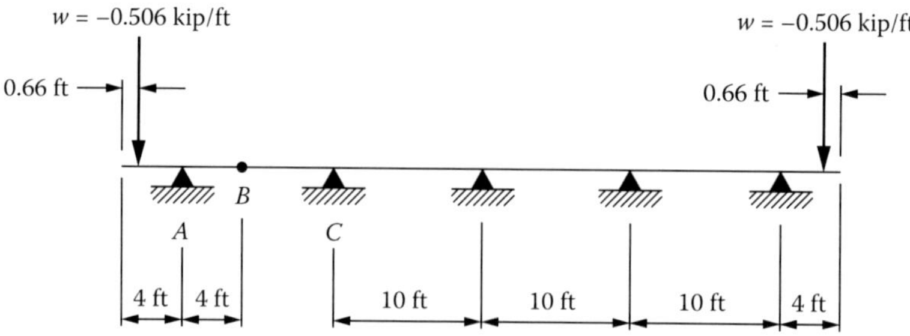

FIGURE 2.54
Moments and reaction for curb and parapet loads.

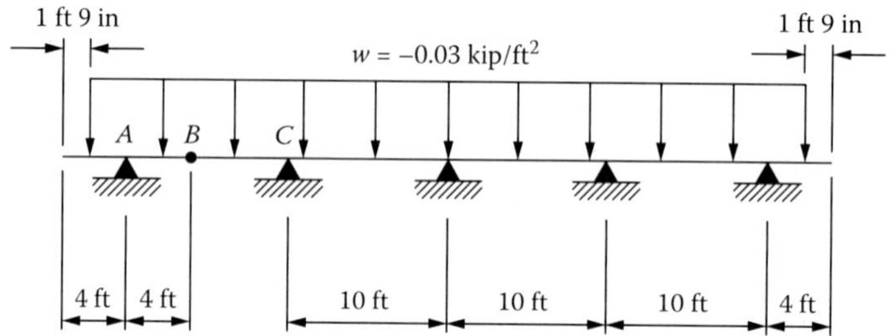

FIGURE 2.55
Moments and reaction for wearing surface loads.

The curb and parapet moments and reaction are as follows and as shown in Figure 2.54.

$$M_A = -1.689 \text{ kip-ft/ft}$$

$$M_B = -0.821 \text{ kip-ft/ft}$$

$$M_C = +0.483 \text{ kip-ft/ft}$$

$$R_A = 0.723 \text{ kips}$$

The future wearing surface moments and reaction are as follows and as shown in Figure 2.55.

$$M_A = -0.076 \text{ kip-ft/ft}$$

$$M_B = +0.195 \text{ kip-ft/ft}$$

$$M_C = -0.300 \text{ kip-ft/ft}$$

$$R_A = 0.195 \text{ kips}$$

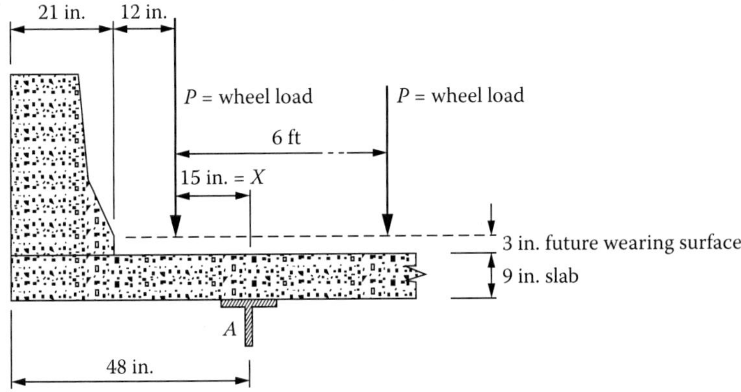

FIGURE 2.56
Live load placement for maximum negative moment.

Step 4: Find the Live Load Force Effects

Where the approximate strip method is used to analyze decks with the slab primarily in the transverse direction, only the axles of the design truck shall be applied to the deck slab.

A Art. 3.6.1.3.1; A Art. 3.6.1.3.3

Wheel loads on the axle are equal and transversely spaced 6.0 ft apart.

A Fig. 3.6.1.2.2-1

The design truck shall be positioned transversely to find extreme force effects such that the center of any wheel load is not closer than 1.0 ft from the face of the curb for the design of the deck overhang and 2.0 ft from the edge of the design lane for the design of all other components.

A Art. 3.6.1.3.1

Find the distance from the wheel load to the point of support, X, where S is the spacing of supporting components using Figure 2.56.

For this example, force effects are calculated conservatively using concentrated wheel loads.

A Art. 4.6.2.1.6

Generally the number of design lanes to be considered across a transverse strip, N_L, should be determined by taking the integer part of the ratio w/12, where w is the clear roadway width in ft and 12 ft is the width of the design lane.

A Art. 3.6.1.1.1

For this example the number of design lanes is

$$N_L = \frac{44.5\text{ ft}}{12\dfrac{\text{ft}}{\text{lane}}} = 3.7\text{ lanes (3 lanes)}$$

The multiple presence factor, m, is 1.2 for one loaded lane, 1.0 for two loaded lanes, and 0.85 for three loaded lanes. (Entries of greater than 1 in [A Tbl. 3.6.1.1.2-1] result from statistical calibration on the basis of pairs of vehicles instead of a single vehicle.)

A Art. 3.6.1.1.2; Tbl. 3.6.1.1.2-1

At this point, the live loads are applied to the deck to find once again the resulting reaction at point A and moments at points A, B, and C.

Find the maximum negative live load moment for overhang.

The critical placement of a single wheel load is at X = 1.25 ft. See Figure 2.57. The equivalent width of a transverse strip for the overhang is

A Tbl. 4.6.2.1.3-1

$$45.0 + 10.0X = 45.0 + (10.0)\left(\frac{15\text{ in}}{12\text{ in}}\right) = 57.5\text{ in} = 4.79\text{ ft}$$

The multiple presence factor, m, is 1.2 for one loaded lane causing the maximum moment.

Therefore, the negative live load moment for overhang is

$$M_A = \frac{-(1.2)(16.0\text{ kips})(1.25\text{ ft})}{4.79\text{ ft}} = -5.01\text{ ft-kips/ft width}$$

Find the maximum positive live load moment in the first interior span.

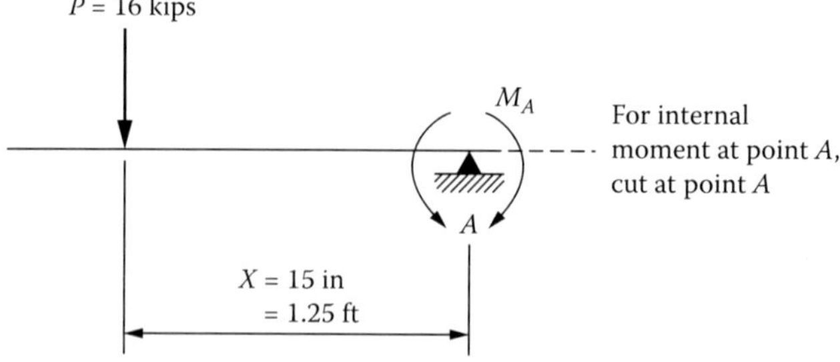

FIGURE 2.57
Live load placement for maximum negative moment, one lane loaded.

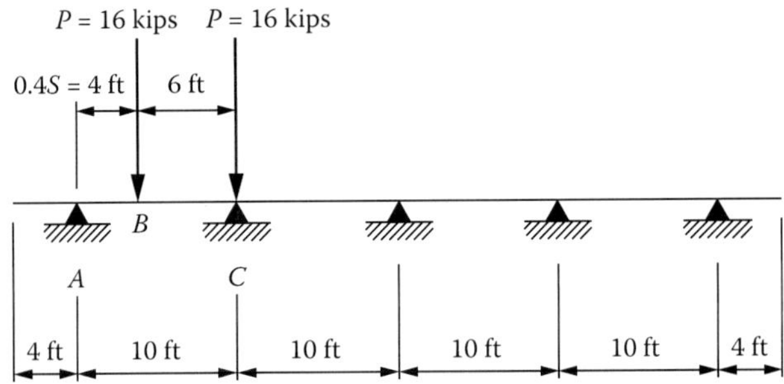

FIGURE 2.58
Live load placement for maximum positive moment in first interior span, one lane loaded.

For a continuous concrete slab with repeating equal spans with all spans loaded with dead loads, the largest positive bending moment occurs at point B, which is 0.4S or 4.0 ft from point A. For this analysis, both single and double lane loadings need to be investigated.

For both cases, the positive moment in the equivalent strip width is

$$26.0 + 6.6S = 26.0 + (6.6)(10\ \text{ft}) = 92\ \text{in} = 7.67\ \text{ft}$$

A Tbl. 4.6.2.1.3-1

To start, a wheel load is located at point B (0.4S), with the other wheel load 6 ft away as dictated by the design truck in the H-93 load model (See AASHTO Art. 3.6.1.2.2). Please see Figure 2.58.

Using structural analysis software or influence line diagrams,

$$R_A = 8.160\ \text{kips}$$

$$M_B = 32.64\ \text{ft-kips}$$

The previous values must be further adjusted to account for strip width and the multiple presence factor, m, of 1.2. The maximum positive live load moment in the first interior span is,

A Art. 3.6.1.1.2

$$R'_A = \frac{(m)(R_A)}{7.67\ \text{ft}} = \frac{(1.2)(8.16\ \text{kips})}{7.67\ \text{ft}} = 1.28\ \text{kips per foot width}$$

$$M'_B = \frac{(m)(M_B)}{7.67\ \text{ft}} = \frac{(1.2)(32.64\ \text{ft-kips})}{7.67\ \text{ft}} = 5.11\ \text{ft-kips per foot width}$$

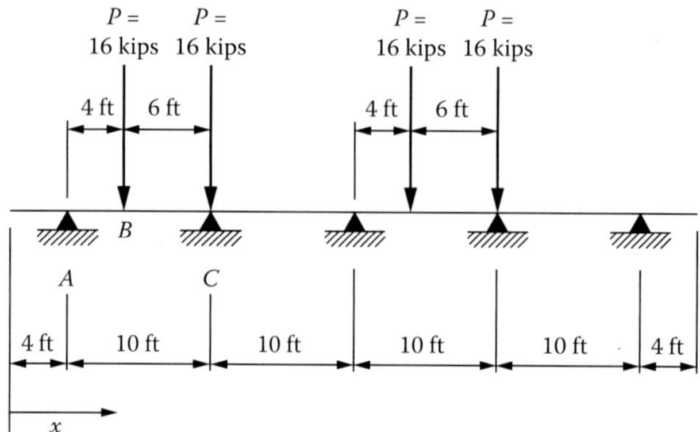

FIGURE 2.59
Live load placement for maximum positive moment, double lane loaded.

If the loadings were placed in the second interior span, they would decrease the effects of the first truck loading in the first interior span.

For double lane loadings, the multiple presence factor, m, is 1.0 at points B and C. Another set of wheel loads is added in the third span, as shown at X = 28 ft and 34 ft from the slab end.

This loading will cause the first interior span moment to increase at B.

For two lanes loaded, see Figure 2.59.

By a structural analysis software program,

$$R_A = 8.503 \text{ kips}$$

$$M_B = 34.01 \text{ ft-kips}$$

Modifying the previous values for strip width and two lanes loaded,

$$R'_S = \frac{(m)(R_A)}{7.67 \text{ ft}} = \frac{(1.0)(8.503 \text{ kips})}{7.67 \text{ ft}} = 1.11 \quad \text{kips per foot width}$$

$$M'_B = \frac{(m)(M_B)}{7.67 \text{ ft}} = \frac{(1.0)(34.01 \text{ ft-kips})}{7.67 \text{ ft}} = 4.43 \ \text{ft-kips per foot width}$$

Thus, the single lane loaded case governs, and the moment effect decreases in the first interior span with increased lane loading cases. Therefore, a scenario of three loaded lanes does not need to be considered for the maximum positive live load moment in the first interior span.

Find the maximum negative live load moment in the first interior span.

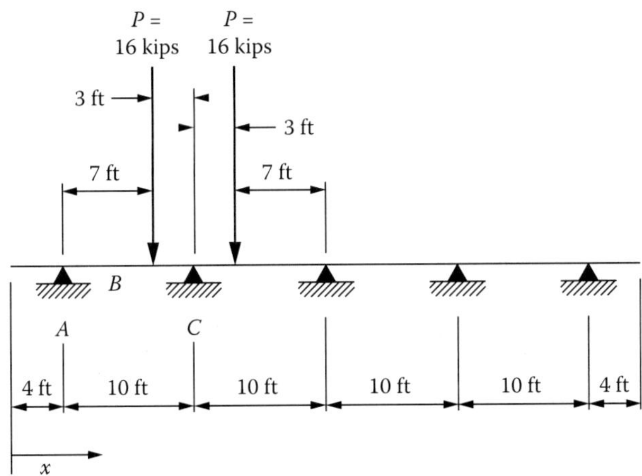

FIGURE 2.60
Live load placement for maximum negative moment in first interior span, one lane loaded.

The critical location for placement of the design truck load for maximum negative moment occurs at the first interior deck support under a one lane load case. Please see Figure 2.60.

The equivalent transverse strip width is

A Tbl. 4.6.2.1.3-1

$$48.0 + 3.0S = 48.0 + (3.0)(10\text{ ft}) = 78\text{ in} = 6.5\text{ ft}$$

Using a structural analysis software program, the maximum negative live load moment at first interior span is,

$M_C = -27.48$ ft-kips

m = 1.2 for one loaded lane

$$M'_C = \frac{(m)(M_C)}{6.5\text{ ft}} = \frac{(1.2)(-27.48\text{ ft-kips})}{6.5\text{ ft}} = -5.07\text{ ft-kips per foot width}$$

The small increase due to the loading of the second truck is not enough to consider because m = 1.0 for two loaded lanes. Therefore, it is only necessary to consider the one lane loaded case.

Find the maximum live load reaction at point A.

The exterior wheel load is placed 1.0 ft from the curb. The width of the strip is conservatively taken as the same as for the overhang. The governing loading is shown in Figure 2.61.

A Art. 3.6.1.3.1

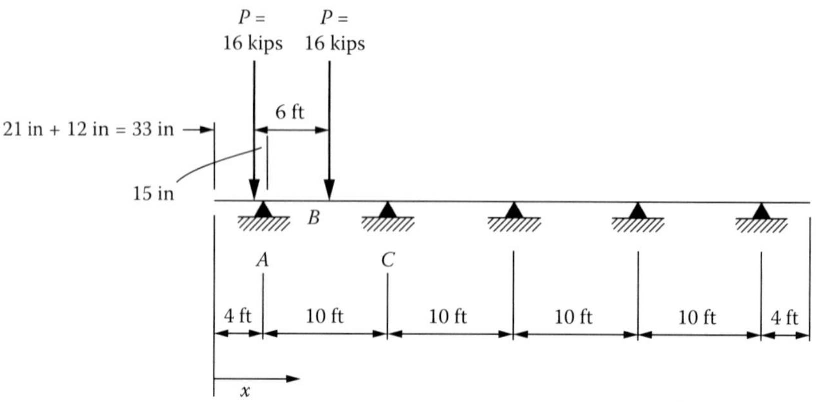

FIGURE 2.61
Live load placement for maximum reaction at first support.

TABLE 2.13
Force Effects Summary Table

	R_A (kips/ft)	M_A (ft-kips/ft)	M_B (ft-kips/ft)	M_C (ft-kips/ft)
Deck slab excluding cantilever	0.444	0.0	0.872	–1.211
Cantilevered slab (overhang)	0.568	–0.904	–0.439	0.258
Curb and parapet	0.723	–1.689	–0.821	0.483
Future wearing surface, FWS	0.195	–0.076	0.195	–0.300
Max negative moment due to overhang	—	–5.010	—	—
Max LL reaction and moment (first span)	1.280	—	5.110	—
Max negative LL moment (first span)	—	—	—	–5.070
Max LL reaction at first support A	6.350	—	—	—

The equivalent strip width of a cast-in-place deck for overhang, where x is the distance from the load to the point of support in feet, is

A Tbl. 4.6.2.1.3-1

$$45.0 + 10.0x = 45.0 + 10.0(1.25 \text{ ft}) = 57.5 \text{ in} = 4.79 \text{ ft}$$

Using a structural analysis software program, when the exterior wheel load is 1.0 ft from the centerline of support, R_A is 24.42 kips. When the exterior wheel load is 1.0 ft from the curb, R_A is 25.36 kips (controls). Please see Table 2.13.

$$R'_A = \frac{mR_A}{\text{equiv strip width}} = \frac{(1.2)(25.36 \text{ kips})}{4.79 \text{ ft}} = 6.35 \text{ kips per foot width}$$

Step 5: Perform the Strength I Limit State Analysis

Each deck component and connection shall satisfy the following equation.

A Art. 1.3.2.1

For loads for which a maximum value of load factor, γ_i, is appropriate,

A Eq. 1.3.2.1-2

$$\eta_i = \eta_D \eta_R \eta_I \geq 0.95$$

For loads for which a minimum value of load factor, γ_i, is appropriate,

A Eq. 1.3.2.1-3

$$\eta_i = \frac{1}{\eta_D \eta_R \eta_I} \leq 1.0$$

$$\eta_D = 1.0$$

A Art. 1.3.3

$$\eta_R = 1.0$$

A Art. 1.3.4

$$\eta_I = 1.0$$

A Art. 1.3.5

The load factor for permanent loads, γ_p, is taken at its maximum value if the force effects are additive and at the minimum value if it subtracts from the dominant force effects. See Table 2.14.

A Tbl. 3.4.1-2

The dynamic load allowance, IM, is 33% of the live-load force effect.

A Art. 3.6.2.1

For the Strength I Limit state, the reaction and moments are,

$$Q = \Sigma \eta_i \gamma_i Q_i = (1.00)\gamma_p DC + (1.00)\gamma_p DW + (1.00)(1.75)(LL + IM)$$

A Tbl. 3.4.1-1, Tbl. 3.4.1-2

TABLE 2.14

Load Factors for Permanent Loads

γ_p	Max	Min
DC	1.25	0.9
DW	1.5	0.65

DW includes the future wearing surface load only, and DC represents all other permanent loads.

R_A = (1.0)(1.25)(0.444 kips/ft + 0.568 kips/ft + 0.723 kips/ft)

+ (1.0)(1.5)(0.195 kips/ft)

+ (1.0)(1.75)(6.35 kips/ft + (0.33)(6.35 kips/ft))

= 17.24 kips per foot width

M_A = (1.0)(1.25)(–0.904 ft-kips/ft + (–1.689 ft-kips/ft) + (1.0)(1.5)(–0.076 ft-kips/ft)

+ (1.0)(1.75)(–5.010 ft-kips/ft + (0.33)(–5.010 ft-kips/ft))

= –15.02 ft-kips per foot width

M_B = (1.0)(1.25)(0.872 ft-kips/ft + (1.0)(0.9)(–0.439 ft-kips/ft + (–0.820 ft-kips/ft)))

+ (1.0)(1.5)(0.195 ft-kips/ft)

+ (1.0)(1.75)(5.110 ft-kips/ft + (0.33)(5.110 ft-kips/ft))

= 12.14 ft-kips per foot width

M_C = (1.0)(1.25)(–1.211 ft-kips/ft + (1.0)(0.9)(0.258 ft-kips/ft + 0.483 ft-kips/ft))

+ (1.0)(1.5)(–0.300 ft-kips/ft)

+ (1.0)(1.75)(–5.070 ft-kips/ft + (0.33)(–5.070 ft-kips/ft))

= –13.10 ft-kips per foot width

It is noted that the load factor for cantilevered slab and curb/parapet load, γ_p, is 0.90 for producing maximum value for M_B and M_C.

A Tbl. 3.4.1-2

For the selection of reinforcement, the negative moments may be reduced to their value at a quarter flange width from the centerline of support. For the purposes of simplicity, this calculation is not performed in this example, but can easily be achieved using classical methods of structural analysis. See Table 2.15.

A Art. 4.6.2.1.6

TABLE 2.15

Strength I Limit State Summary (Factored Values)

Maximum extreme positive moment, M_B	12.14 ft-kips/ft
Maximum extreme negative moment, M_A	−15.02 ft-kips/ft > M_c = −13.10 ft-kips/ft
Maximum reaction, R_A	17.24 kips/ft

Step 6: Design for Moment

The compressive strength of the concrete at 28 days, f'_c, is 4.5 kips/in^2. The specified minimum yield point of the steel, F_y, is 60 kips/in^2. Determine the configuration of epoxy-coated steel reinforcement required.

A Art. 5.10.3.2

The maximum spacing of primary reinforcement for slabs is 1.5 times the thickness of the member or 18.0 in. By using the structural slab thickness of 8.5 in, the maximum spacing of reinforcement becomes:

$$s_{max} = (1.5)(8.5 \text{ in}) = 12.75 \text{ in} < 18 \text{ in [controls]}$$

The required concrete cover is 2.5 in for the unprotected main reinforcement for deck surfaces subject to wear and 1.0 in for the bottom of cast-in-place slabs. For conservatism and simplicity of this example, 2.5 in is used for both top and bottom covers.

A Tbl. 5.12.3-1

Note that if epoxy-coated bars are used, 1.0 in is required for covers for top and bottom.

A Art. 5.12.3

Assuming no. 5 steel rebar,

$$d_b = 0.625 \text{ in}$$

$$A_b = 0.31 \text{ in}^2$$

See Figure 2.62 for the reinforcement placement.

Factored moment M_A = −15.02 ft-kips is equal to the factored flexural resistance M_r.

The distance from the extreme compression fiber to the centroid of reinforcing bars is

$$d = 9.0 \text{ in} - 0.5 \text{ in (integral wearing surface)} - 2.5 \text{ in (cover)} - 0.625 \text{ in}/2 \text{ (bar diameter)}$$

$$= 5.68 \text{ in}$$

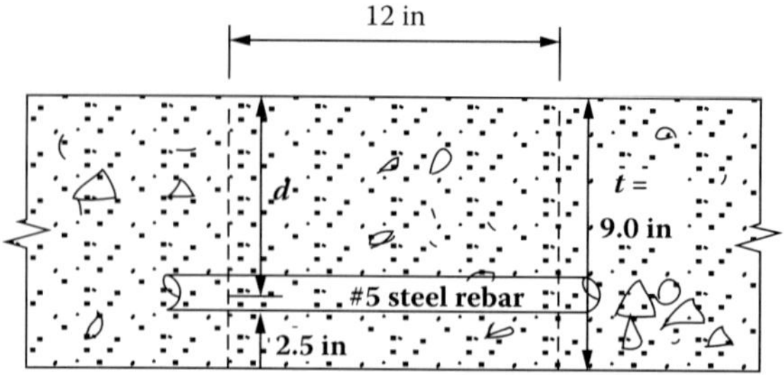

FIGURE 2.62
Deck slab section for reinforcement placement.

The depth of the equivalent stress block is

$$a = \frac{A_s f_y}{0.85\, f'_c\, b} = \frac{A_s(60\text{ ksi})}{(0.85)(4.5\text{ ksi})(12\text{ in})} = 1.307 A_s$$

A Art. 5.7.3.2

The factored resistance M_r is

$$M_r = \Phi M_n = \Phi A_s f_y\left(d - \frac{a}{2}\right)$$

where:
Φ = resistance factor = 0.9 for flexure in reinforced concrete
M_n = nominal resistance

Let M_r equal to the factored moment M_u.

$$M_u\ (= M_A) = M_r$$

$$\left(15.02\frac{\text{ft-kips}}{\text{ft}}\right)\left(12\frac{\text{in}}{\text{ft}}\right) = 0.9\, A_s(60\text{ ksi})\left(5.68\text{ in} - \left(\frac{1.307}{2}A_s\right)\right)$$

A Art. 5.5.4.2.1

The minimum area of steel needed is

$$A_s = 0.64\text{ in}^2/\text{ft of slab width}$$

Use no. 5 bars spaced at 5.5 in. So, the provided area of steel is

$$A_s = A_b\left(\frac{12\text{ in width}}{\text{spacing}}\right) = 0.31\text{ in}^2\left(\frac{12\text{ in}}{5.5\text{ in}}\right) = 0.68\frac{\text{in}^2}{\text{ft}} > A_s = 0.64\text{ in}^2/\text{ft [good]}$$

Check the moment capacity,

$$a = \frac{A_s f_y}{0.85\, f'_c\, b} = \frac{(0.68\text{ in}^2)(60\text{ ksi})}{(0.85)(4.5\text{ ksi})(12\text{ in})} = 0.889\text{ in}$$

Confirm that M_r is equal to or greater than the factored moment M_A (= M_u)

$$M_r = \varphi M_n = \varphi A_s f_y \left(d - \frac{a}{2} \right)$$

$$= (0.9)(0.68\ \text{in}^2)(60\ \text{ksi})\left(5.68\ \text{in} - \frac{0.889\ \text{in}}{2} \right)\left(\frac{1\ \text{ft}}{12\ \text{in}} \right)$$

$$= 16.0\ \text{ft-kips/ft} > M_u = 15.02\ \text{ft-kips/ft [OK]}$$

Check minimum steel.

The minimum reinforcement for flexural components is satisfied if a factored flexural resistance $\Phi M_n = M_r$ is at least equal to the lesser of 1.2 times the cracking moment, M_{cr}, and 1.33 times the factored moment required by the applicable strength load combination.

A Art. 5.7.3.3.2

Where slabs are designed for a noncomposite section to resist all loads, the cracking moment is,

A Eq. 5.7.3.3.2-1

$$M_{cr} = S_{nc} f_r$$

where,
S_{nc} = section modulus of the noncomposite section
f_r = modulus of rupture

For normal weight concrete, the modulus of rupture of concrete is

A Art. 5.4.2.6

$$f_r = 0.37\sqrt{f'_c} = 0.37\sqrt{4.5\ \text{kips/in}^2} = 0.785\ \text{kips/in}^2$$

The section modulus for the extreme fiber of the noncomposite section where tensile stress is caused by external loads is

$$S_{nc} = \left(\frac{1}{6} \right)(12\ \text{in})(8.5\ \text{in})^2 = 144.5\ \text{in}^3$$

The cracking moment is

$$M_{cr} = S_{nc}f_r = (144.5\text{ in}^3)\left(0.785\frac{\text{kips}}{\text{in}^2}\right) = 113.4\text{ in-kips}$$

$$1.2\,M_{cr} = (1.2)(113.4\text{ in-kips})\left(\frac{1\text{ ft}}{12\text{ in}}\right)$$

$$= 11.34\text{ ft-kips/ft}$$

$$M_u = M_A$$

$$1.33\,M_u = (1.33)\left(15.02\frac{\text{ft-kips}}{\text{ft}}\right) = 19.98\text{ ft-kips/ft}$$

$$1.2\,M_{cr} = 11.34\text{ ft-kips/ft [controls]}$$

$$11.34\text{ ft-kips/ft} < \Phi M_n = M_r = 16.0\text{ ft-kips/ft [OK]}$$

Reinforcement transverse to the main steel reinforcement (which is perpendicular to traffic) is placed in the bottom of all slabs. The amount shall be a percentage of the main reinforcement required as determined in the following formula.

A Art. 9.7.3.2

For primary reinforcement perpendicular to traffic, S_e is the effective span length; the following must be true.

A Art. 9.7.3.2

$$\frac{220}{\sqrt{S_e}} \le 67\%$$

For slabs supported on steel girders, S_e is the distance between flange tips, plus the flange overhang, taken as the distance from the extreme flange tip to the face of the web, disregarding any fillet. (For a W12 × 65, $t_w = 0.39$ in).

A Art. 9.7.2.3

$$S_e = 10\text{ ft} - (0.39\text{ in})\left(\frac{1\text{ ft}}{12\text{ in}}\right) = 9.97\text{ ft}$$

$$S_e \cong 10\text{ ft}$$

Reinforcement shall be placed in the secondary direction in the bottom of slab as a percentage of the primary reinforcement perpendicular to traffic for positive moment as follows:

A Art. 9.7.3.2

$$\frac{220}{\sqrt{S_e}} \le 67\%$$

$$\frac{220}{\sqrt{S_e}} = \frac{220}{\sqrt{10\text{ ft}}} = 69.6\% > (67\%)\ \text{[no good]}$$

Therefore, use 67%.

$$A_s = 0.67\ A_s = (0.67)(0.68\text{ in}^2) = 0.46\text{ in}^2/\text{ft}$$

For longitudinal bottom bars, use no. 5 ($A_b = 0.31\text{ in}^2$) at 8 in,

$$A_s = A_b\left(\frac{12\text{ in width}}{\text{spacing}}\right) = \left(0.31\text{ in}^2\right)\left(\frac{12\text{ in}}{8\text{ in}}\right)$$

$$= 0.46\text{ in}^2/\text{ft}$$

The reinforcement needed in each direction for the shrinkage and temperature reinforcement shall be

A Art. 5.10.8

$$A_{s,temp} \ge \frac{1.30\,bh}{2(b+h)f_y}$$

where:
b = least width of component section (12 in)
h = least thickness of component section (8.5 in)
f_y = specified yield strength of reinforcing bars ≤ 75 kips/in²

A Eq. 5.10.8-1

$$A_{s,temp} \ge \frac{1.30(12\text{ in})(8.5\text{ in})}{2(12\text{ in}+8.5\text{ in})\left(60\dfrac{\text{kips}}{\text{in}^2}\right)} = 0.054\frac{\text{in}^2}{\text{ft}}$$

$0.11 \le A_{s,temp} \le 0.60$, so use #3 bars ($A_b = 0.11\text{ in}^2/\text{ft}$)

A Eq. 5.10.8-2

The primary and secondary reinforcement already selected provide more than this amount; however, for members greater than 6.0 in thickness the shrinkage and temperature reinforcement is to be distributed equally on both faces. The maximum spacing of this reinforcement is 3.0 times the slab thickness or 18 in. For the top face longitudinal bars, the area of the temperature reinforcement, $A_{s,temp}$, is 0.11 in^2/ft. Use no. 4 bars at 18 in, providing A_s = 0.13 in^2/ft.

A Eq. 5.10.8

For primary reinforcement, use no. 5 bars at 5.5 in. For longitudinal bottom bars, use no. 5 at 8 in. For longitudinal top bars use no. 4 at 18 in.

Step 7: Alternate Solution Utilizing Empirical Design Method

The empirical method is a simplified design method that can be utilized when provisions of Art. 9.7.2 are satisfied. This method does not require extensive structural analysis (by the moment distribution method, influence line aids, or computer methods) as required in the previously worked solution. As such, the empirical method may prove useful in written testing situations.

Empirical design shall not be applied to deck overhangs.

A Art. 9.7.2.2

Verify that requirements for the empirical design method use are satisfied.

A Art. 9.7.2.4

Find the effective length of the slab cut out or:

A Art. 9.7.2.3

For a W 12 × 65 steel girder, the web thickness, t_w, is 0.39 in.

$$L_{effective} = 10\text{ ft} - t_w = 10\text{ ft} - (0.39\text{ in})\left(\frac{1\text{ ft}}{12\text{ in}}\right) = 9.97\text{ ft}$$

Effective length of slab cannot exceed 13.5 ft. [OK]

A Art. 9.7.2.4

Ratio of effective length to slab depth cannot exceed 18.0 and is not less than 6.0.

A Art. 9.7.2.4

The slab depth, d, is

$$d = (9\text{ in})\left(\frac{1\text{ ft}}{12\text{ in}}\right) = 0.75\text{ ft}$$

$$\frac{L_{effective}}{d} = \frac{9.94\text{ ft}}{0.75\text{ ft}} = 13.3\text{ [OK]}$$

Slab thickness is not less than 7.0 in excluding wearing surface. [OK]

A Art. 9.7.2.4

Overhang beyond center line of outside girder at least equals 5.0 times depth of slab.

A Art. 9.7.2.4

$$4\text{ ft} > (5.0)(0.75\text{ ft}) = 3.75\text{ ft [OK]}$$

Slab core depth is not less than 4.0 in. [OK]

A Art. 9.7.2.4

The following conditions are also met or assumed to be satisfied:

A Art. 9.7.2.4

- Cross frames or diaphragms are used throughout.
- Deck is uniform depth, excluding haunches.
- Supporting components are made of steel.
- Concrete strength, f'_c, is at least 4000 psi.
- Deck is made composite with supporting structural components (in both positive and negative moment regions, with adequate shear connectors).

Four layers of isotropic reinforcement are required.

A Art. 9.7.2.5

Reinforcement shall be closest to outside surface as possible.

A Art. 9.7.2.5

Outermost reinforcing layer shall be in effective direction.

A Art. 9.7.2.5

Spacing shall not exceed 18.0 in.

A Art. 9.7.2.5

Reinforcing shall be Grade 60 steel or better.

A Art. 9.7.2.5

Design top reinforcing steel:

For top layer, 0.18 in^2/ft is required in each direction.

A Art. 9.7.2.5

Try #4 bars with 12 in spacing.
Total area of steel is equal to 0.198 in^2.
At 12 in spacing, we have 0.198 in^2 per foot.
For bottom layer, 0.27 in^2/ft is required in each direction.

A Art. 9.7.2.5

Try #5 bars spaced at 12 in.
Total area of steel = 0.309 in^2.
At 12 in spacing, we have 0.309 in^2.
For bottom reinforcing, use #5 spaced at 12 in in each direction.

It can be seen that significantly less reinforcing bar is required when designed by the empirical method. This is due to a complex arching affect that governs behavior of the slab in these conditions.

A Art. C9.7.2.1

Design Example 6: Prestressed Interior Concrete Girder

Situation

An interior prestressed concrete girder for a two-lane simply supported highway bridge in central New York State is to be designed. The spacing of the bridge's five girders is 7 ft 6 in. The width of the exterior beam overhang is 3 ft 9 in.

The design load is AASHTO HL-93. Allow for a future wearing surface, FWS, of 3 in bituminous concrete with a load, w_{FWS}, of 0.140 $kips/ft^3$, and use Strength I load combination for load resistance factor design.

L	bridge span	80 ft
	integral wearing surface of slab	0.5 in
W_{FWS}	load of future wearing surface of 3 in bituminous pavement	0.140 $kips/ft^3$

A_{ps}	area of grade 270 prestressing steel (44 strands at ½ in diameter; 7 wire = (44)(0.153 in²))	6.732 in²
E_s	modulus of elasticity of prestressing steel	28500 ksi
$f'_{cg} = f'_c$	compressive strength of concrete at 28 days for prestressed I-beams	6500 lbf/in²
f'_{ci}	compressive strength of concrete at time of initial prestress	6000 lbf/in²
f'_{cs}	compressive strength of concrete for 8 in slab	4500 lbf/in²
f_{pu}	specified tensile strength of prestressing steel	270 kips/in²

The basic beam properties are as follows:

A_g	cross-sectional area of basic beam	762 in²
I_g	moment of inertia of basic beam about centroidal axis, neglecting reinforcement	212,450 in⁴
$S_{nc,top}$	top section modulus for the extreme fiber of the noncomposite section	7692 in³
$S_{nc,bottom}$	bottom section modulus for the extreme fiber of the noncomposite section	9087 in³
y_b	distance from the bottom fiber to the centroid of the basic beam	23.38 in

Please see Figure 2.63.

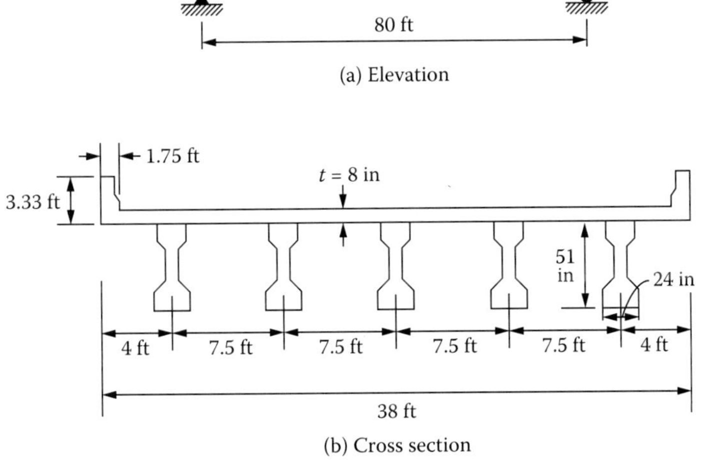

FIGURE 2.63
Prestressed concrete interior girder design example.

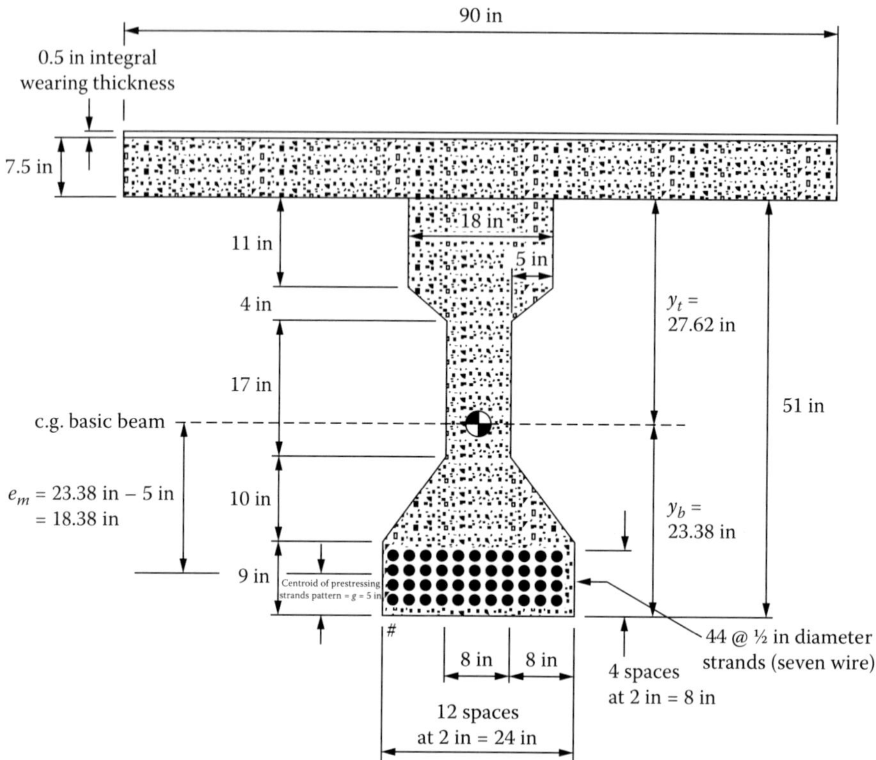

FIGURE 2.64
Cross section of girder with composite deck.

Requirements

Determine the composite section properties; the factored design moment for Strength I Limit State at midspan, M_u; and the girder moment capacity, ΦM_n (= M_r). For load combination Limit State Service I, determine the concrete stresses in the girder at midspan at release of prestress and the final concrete stresses after all losses (except friction) at midspan. Please see Figure 2.64.

Step 1: Determine the Composite Section Properties

The thickness of the following sections of precast concrete beams must meet the following specifications:

A Art. 5.14.1.2.2

top flange > 2 in [OK]
web > 5 in [OK]
bottom flange > 5.0 in [OK]

Calculate the minimum depth including deck for precast prestressed concrete I-beams with simple spans, d_{min}.

A Tbl. 2.5.2.6.3-1

$$d_{min} = 0.045\ L = (0.045)(80\ \text{ft})(12\ \text{in/ft}) = 43.2\ \text{in}$$

$$d = 7.5\ \text{in} + 51\ \text{in} = 58.5\ \text{in} > 43.2\ \text{in [OK]}$$

b_{top}	top flange width	18 in
L	effective span length (actual span length)	80 ft
t_s	slab thickness	7.5 in

The effective slab flange width for interior beams is

A Art. 4.6.2.6.1

$$b_i = b_e = S = (7.5\ \text{ft})(12\ \text{in/ft}) = 90\ \text{in}$$

The normal unit weight of concrete, w_c, is 0.15 kips/ft³.

A Tbl. 3.5.1-1

The modulus of elasticity of concrete at 28 days for prestressed I-beams is

A Eq. 5.4.2.4-1

$$E_{cg} = 33{,}000\ w_c^{1.5}\sqrt{f'_{cg}}$$

$$= (33{,}000)\left(0.15\frac{\text{kips}}{\text{ft}^3}\right)^{1.5}\sqrt{6.5\frac{\text{kips}}{\text{in}^2}}$$

$$= 4890\ \text{kips/in}^2$$

The modulus of elasticity of the slab is

$$E_s = 33{,}000\ w_c^{1.5}\sqrt{f'_{cs}}$$

$$= (33{,}000)\left(0.15\frac{\text{kips}}{\text{ft}^3}\right)^{1.5}\sqrt{4.5\frac{\text{kips}}{\text{in}^2}}$$

$$= 4070\ \text{kips/in}^2$$

The modular ratio is

$$n = \frac{E_{cg}}{E_{cs}} = \frac{4890\frac{\text{kips}}{\text{in}^2}}{4070\frac{\text{kips}}{\text{in}^2}} = 1.2$$

The transformed area of the slab is

$$(7.5\text{ in})\left(\frac{90\text{ in}}{1.2}\right) = 562.5\text{ in}^2$$

Please see Figure 2.65.

The area of the composite section, A_c, is

$$A_c = A_g + \text{transformed area of slab}$$

$$= 762\text{ in}^2 + 562.5\text{ in}^2$$

$$= 1324.5\text{ in}^2$$

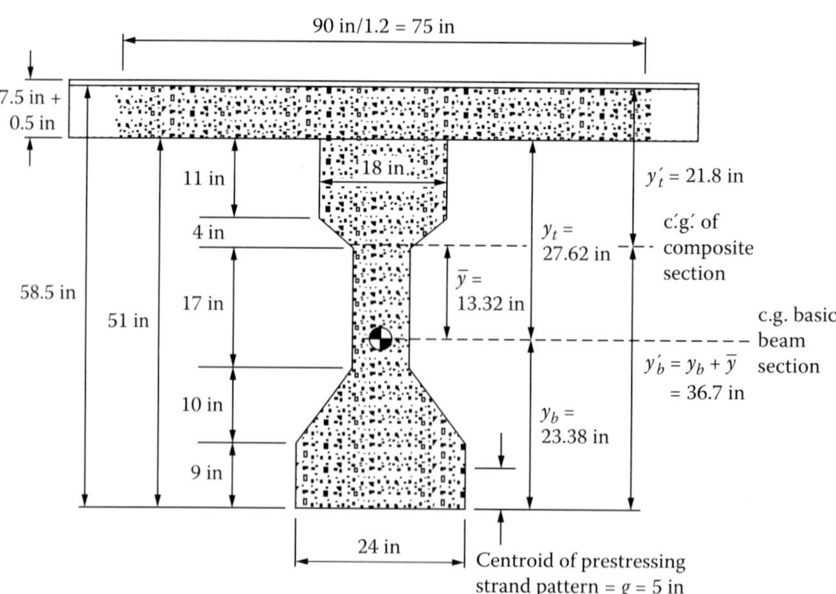

FIGURE 2.65
Area transformed section of girder section.

From the centroid of the basic beam section, c.g., solve for $\bar{y}$.

$$\left(1324.5 \text{ in}^2\right)\bar{y} = \left(562.5 \text{ in}^2\right)\left(y_t + \frac{7.5 \text{ in}}{2}\right)$$

$$\bar{y} = \frac{\left(562.5 \text{ in}^2\right)\left(27.62 \text{ in} + 3.75 \text{ in}\right)}{1324.5 \text{ in}^2}$$

$$y'_b = y_b + \bar{y} = 23.38 \text{ in} + 13.32 \text{ in} = 36.7 \text{ in}$$

$$y'_t = \left(51 \text{ in} + 7.5 \text{ in}\right) - y'_b = 21.8 \text{ in}$$

The composite moment of inertia, I_c, is

$$I_c = I_g + A_g\bar{y}^2 + \frac{(75 \text{ in})(7.5 \text{ in})^3}{12} + \left(562.5 \text{ in}^2\right)\left(y'_t - \frac{7.5 \text{ in}}{2}\right)^2$$

$= 212{,}450 \text{ in}^4 + (762 \text{ in}^2)(13.32 \text{ in})^2 + 2636.7 \text{ in}^4 + (562.5 \text{ in}^2)(21.8 \text{ in} - 3.75 \text{ in})^2$

$= 533{,}546.5 \text{ in}^4$

For the bottom extreme fiber, the composite section modulus is

$$S_{bc} = \frac{I_c}{y'_b} = \frac{533{,}546.5 \text{ in}^4}{36.7 \text{ in}} = 14{,}538.0 \text{ in}^3$$

For the top extreme fiber (slab top), the composite section modulus is (see Table 2.16),

$$S_{tc} = \frac{I_c}{y'_t} = \frac{533{,}546.5 \text{ in}^4}{21.8 \text{ in}} = 24{,}474.6 \text{ in}^3$$

TABLE 2.16

Summary of Section Properties

Basic		**Composite**	
A_g	762 in^2	A_c	1324.5 in^2
y_t	27.62 in	y'_t	21.8 in
y_b	23.38 in	y'_b	36.7 in
I_g	212,450 in^4	I_c	533,546.5 in^4
S_{nct}	7692 in^3	S_{tc}	24,474.6 in^3
S_{ncb}	9087 in^3	S_{bc}	14,538 in^3

Step 2: Determine the Factored Maximum Moment

Use AASHTO HL-93 live load model to find the unfactored moment and shear due to live load.

For all components excluding deck joints and fatigue, the dynamic load allowance, IM, is 33%.

A Art. 3.6.2.1

The beam spacing, S is 7.5 ft.

The bridge span, L, is 80 ft.

Calculate the distribution factor for moments, DFM_{si}, for interior beams with one lane loaded.

A Tbl. 4.6.2.2.2b-1 or Appendix A

NOTE: The multiple presence factor, m, is already included in the approximate equations for live load distribution factors.

A Art. C3.6.1.1.2

$$\left(\frac{K_g}{12\,Lt_s^3}\right)^{0.1} = 1.09 \text{ for the cross-section type (k).}$$

A Tbls. 4.6.2.2.1-1; 4.6.2.2.1-2; 4.6.2.2.2b-1 or Appendix A

$$DFM_{si} = \left(0.06 + \left(\frac{S}{14}\right)^{0.4}\left(\frac{S}{L}\right)^{0.3}\left(\frac{K_g}{12\,Lt_s^3}\right)^{0.1}\right)$$

$$= \left(0.06 + \left(\frac{7.5\text{ ft}}{14}\right)^{0.4}\left(\frac{7.5\text{ ft}}{80\text{ ft}}\right)^{0.3}(1.09)\right) = 0.477$$

Calculate the distribution factor for moments, DFM_{mi}, for interior beams with two lanes loaded.

$$DFM_{mi} = \left(0.075 + \left(\frac{S}{9.5}\right)^{0.6}\left(\frac{S}{L}\right)^{0.2}\left(\frac{K_g}{12\,Lt_s^3}\right)^{0.1}\right)$$

$$= \left(0.075 + \left(\frac{7.5\text{ ft}}{9.5}\right)^{0.6}\left(\frac{7.5\text{ ft}}{80\text{ ft}}\right)^{0.2}(1.09)\right) = 0.664 \text{ [controls]}$$

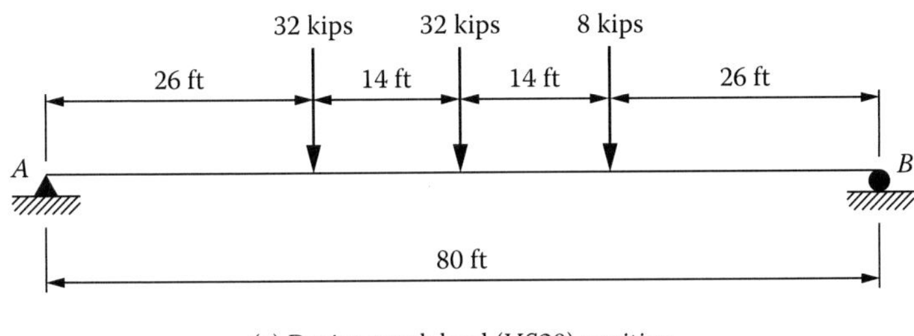

(a) Design truck load (HS20) position

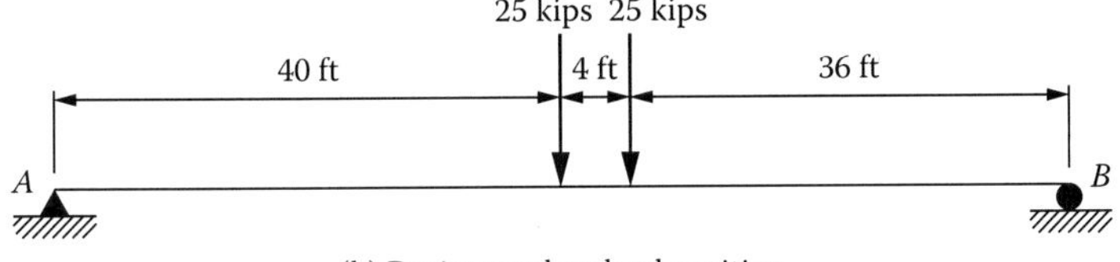

(b) Design tandem load position

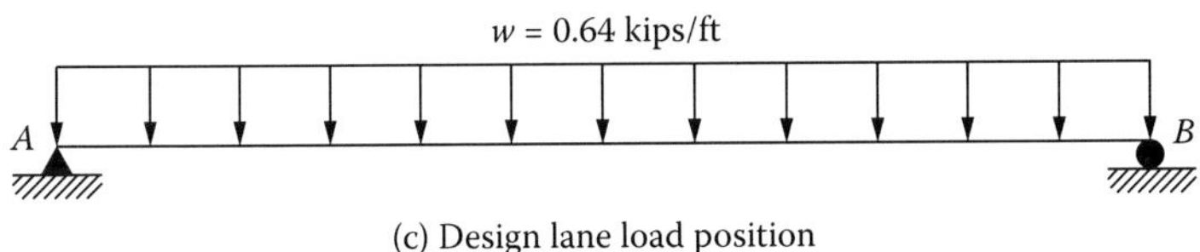

(c) Design lane load position

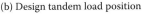

FIGURE 2.66
Bending moments at midspan due to HL-93 loading.

Approximate maximum bending moments at midspan due to HL-93 loading as follows and shown in Figure 2.66.

Find the truck load (HS-20) moment, M_{tr}

$$\Sigma M_B = 0$$

$$0 = 8\text{ kips}(26\text{ ft}) + 32\text{ kips}(40\text{ ft}) + 32\text{ kips}(54\text{ ft}) - R_A(80\text{ ft})$$

$$R_A = 40.2\text{ kips}$$

$$M_{tr} = (40.2\text{ kips})(40\text{ ft}) - 32\text{ kips}(14\text{ ft})$$

$$= 1160\text{ ft-kips}$$

Find the tandem load moment M_{tandem}

$$\Sigma M_B = 0$$

$$0 = 25\text{ kips}(36\text{ ft}) + 25\text{ kips}(40\text{ ft}) - R_A(80\text{ ft})$$

$$R_A = 23.75 \text{ kips}$$

$$M_{tandem} = (23.75 \text{ kips})(40 \text{ ft}) = 950 \text{ ft-kips}$$

Find the lane load moment, M_{ln}

$$M_{ln} = \frac{\left(0.64\frac{\text{kips}}{\text{ft}}\right)\left(80\frac{\text{kips}}{\text{ft}}\right)^2}{8} = 512 \text{ ft-kips}$$

The maximum live load plus impact moment per girder is defined by the following equation:

$$M_{LL+IM} = DFM_{mi}\left(\left(M_{tr} \quad \text{or} \quad M_{tandem}\right)\left(1+\frac{IM}{100}\right)+M_{ln}\right)$$

$$= (0.664)((1160 \text{ ft-kips})(1 + 0.33) + 512 \text{ ft-kips})$$

$$= 1364.38 \text{ ft-kips per girder}$$

Find the moment due to DC, M_{DC}. The beam weight is

$$w = \left(762 \text{ in}^2\right)\left(\frac{1 \text{ ft}^2}{(12 \text{ in})(12 \text{ in})}\right)\left(0.15\frac{\text{kips}}{\text{ft}^3}\right)$$

$$= 0.79 \text{ kips/ft}$$

The moment due to the beam weight is

$$M_g = \left(\frac{wL^2}{8}\right) = \frac{\left(0.79\frac{\text{kips}}{\text{ft}}\right)(80 \text{ ft})^2}{8} = 632 \text{ ft-kips}$$

The slab dead load is

$$w = (8 \text{ in})\left(\frac{1 \text{ ft}}{(12 \text{ in})}\right)\left(0.15\frac{\text{kips}}{\text{ft}^3}\right)(7.5 \text{ ft})$$

$$= 0.75 \text{ kips/ft}$$

The moment due to the slab dead load is

$$M_D = \left(\frac{wL^2}{8}\right) = \frac{\left(0.75\frac{\text{kips}}{\text{ft}}\right)(80\text{ ft})^2}{8} = 600 \text{ ft-kips}$$

The total moment due to DC is

$$M_{DC} = M_g + M_D = 632 \text{ ft-kips} + 600 \text{ ft-kips} = 1232 \text{ ft-kips}$$

The future wearing surface is 3 in.
Find the moment due to DW, M_{DW}.

$$w_{DW} = \left(\frac{3.0 \text{ in}}{12\frac{\text{in}}{\text{ft}}}\right)\left(0.140\frac{\text{kips}}{\text{ft}^3}\right)(7.5 \text{ ft}) = 0.263 \text{ kips/ft}$$

The moment due to DW is (also see Table 2.17)

$$M_{DW} = \left(\frac{w_{DW}L^2}{8}\right) = \frac{\left(0.263\frac{\text{kips}}{\text{ft}}\right)(80\text{ ft})^2}{8} = 210.4 \text{ ft-kips}$$

Limit States

A Art. 1.3.2

Find load modifier η_i.

$$\eta_i = \eta_D\eta_R\eta_I \geq 0.95$$

η_D ductility factor 1.00 for conventional design

A Art. 1.3.3

TABLE 2.17

Unfactored Moments per Girder

Load Type	Moment (ft-kips)
DC	1232.0
DW	210.4
LL + IM	1364.38

η_R redundancy factor 1.00 for conventional levels of redundancy

A Art. 1.3.4

η_I operational importance factor 1.00 for typical bridges

A Art. 1.3.5

Find the factored maximum moment for the Strength I Limit State

A Tbls. 3.4.1-1; 3.4.1-2

$Q = \Sigma\eta_i\ \gamma_i\ Q_i$

$Q = 1.0[\gamma_p\ DC\ \gamma_p\ DW + 1.75\ (LL + IM)]$

$M_u = (1.0)(1.25)(1232.0\ \text{ft-kips}) + (1.50)(210.4\ \text{ft-kips}) + (1.75)(1364.38\ \text{ft-kips})$

$= 4243.26\ \text{ft-kips}$

Step 3: Determine the Girder Moment Capacity, ΦM_n (= M_r)

Find the average stress in prestressing steel.

Assume rectangular behavior.

Find the distance from the extreme compression fiber to the neutral axis, c.

A_{ps}	area of the prestressing steel	6.732 in^2
b	width of the compression face member	90.0 in
d_p	distance from extreme compression fiber to the centroid of the prestressing tendons	in
f'_c	compressive strength of concrete at 28 days	6.5 kips/in^2
f_{ps}	average stress in prestressing steel at the time for which the nominal resistance of member is required	kips/in^2
f_{pu}	specified tensile strength of prestressing steel	270 kips/in^2
f_s	stress in the tension reinforcement	0 kips/in^2
f'_s	stress in the compression reinforcement	0 kips/in^2
k	shear-buckling coefficient for webs	0.28

A Tbl. C5.7.3.1.1

t_s	slab thickness	8.0 in

The distance from the compression fiber to the centroid of prestressing tendons is

$$d_p = (51\text{ in} - 5\text{ in}) + 8\text{ in} = 54\text{ in}$$

The factor for concrete strength is

A Art. 5.7.2.2

$$\beta_1 = 0.85 - (0.05)(f'_c - 4\text{ ksi}) = (0.05)(6.5\text{ ksi} - 4\text{ ksi})$$
$$= 0.725$$

The distance from the extreme compression fiber to the neutral axis is

A Eq. 5.7.3.1.1-4

$$c = \frac{A_{ps}f_{pu} + A_s f_s - F'_s f'_s}{0.85\, f'_c\, \beta_1 b + kA_{ps}\dfrac{f_{ps}}{d_p}}$$

$$= \frac{(6.732\text{ in}^2)(270\text{ ksi}) + (0\text{ in}^2)(0\text{ ksi}) - (0\text{ ksi})(0\text{ ksi})}{(0.85)(6.5\text{ ksi})(0.725)(90\text{ in}) + (0.28)(6.732\text{ in}^2)\dfrac{270\text{ ksi}}{54.0\text{ in}}}$$

$$= 4.91\text{ in} < t_s = 8\text{ in [assumption OK]}$$

The average stress in the prestressing steel is

A Eq. 5.7.3.1.1-1

$$f_{ps} = f_{pu}\left(1 - k\frac{c}{d_p}\right)$$

$$= (270\text{ ksi})\left(1 - (0.28)\left(\frac{4.91\text{ in}}{54\text{ in}}\right)\right) = 263.1\text{ ksi}$$

Find the flanged section factored flexural resistance

A Art. 5.7.3.2.2

$$a = \beta_1 c = (0.725)(4.91\text{ in}) = 3.56\text{ in}$$

a	depth of equivalent rectangular stress block	3.56 in
Φ	resistance factor	1.00

A Art. 5.5.4.2

The nominal flexural resistance, M_n, and the factored resistance, M_r, are,

A Art. 5.7.3.2

Neglecting nonprestressed reinforcement,

$$M_n = A_{ps}f_{ps}\left(d_p - \frac{a}{2}\right)$$

$$M_n = \left(6.732\ \text{in}^2\right)\left(263.7\ \text{ksi}\right)\left(54\ \text{in} - \frac{3.56\ \text{in}}{2}\right) = 7725.20\ \text{ft-kips}$$

$$M_r = \Phi M_n = (1.0)(7725.20\ \text{ft-kips}) > M_u$$

$$= 4243.46\ \text{ft-kips for Strength I Limit State [OK]}$$

Step 4: Determine Concrete Stresses at Midspan at Release of Prestress

Temporary allowable concrete stresses before losses (at time of initial prestress) due to creep and shrinkage are as follows. See Figure 2.67.

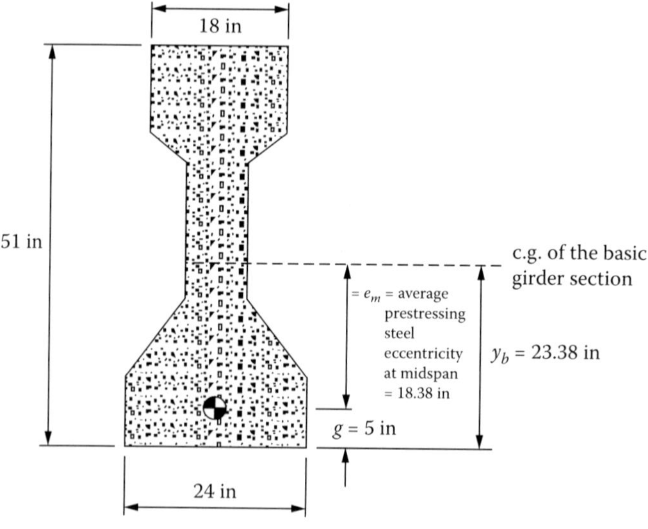

FIGURE 2.67
Girder I-beam section.

For compression,

A Art. 5.9.4.1.1

$$f_{ci} = 0.6\, f'_{ci} = (0.6)(6.0\text{ ksi}) = 3.6\text{ ksi}$$

For tension,

A Art. 5.9.4.1.2

$$f_{ti} = 0.24\sqrt{f'_{ci}} = 0.24\sqrt{5.5\text{ ksi}} = 0.563\text{ ksi}$$

Find the reduced tendon stress immediately after transfer due to elastic shortening.

A_g	gross area of cross section	762 in^2
A_{ps}	area of prestressing steel	6.732 in^2
e_m	average prestressing steel eccentricity at midspan	18.38 in

$$E_{ci} = 33{,}000\, w_c^{1.5}\sqrt{f'_{ci}} = 33{,}000\left(0.15\,\frac{\text{kips}}{\text{ft}^3}\right)^{1.5}\sqrt{6.0\text{ ksi}} = 4696\text{ ksi}$$

A Eq. 5.4.2.4-1

E_{ci}	modulus of elasticity of concrete at transfer	4696 ksi
E_p	modulus of elasticity of prestressing tendon	28,500 ksi
f_{cgp}	concrete stress at center of gravity of prestressing tendons due to prestressing force immediately after transfer and self-weight of member at section of maximum moment	ksi
f_{pbt}	stress immediately prior to transfer	ksi
f_{pj}	stress at jacking	ksi
f_{pu}	specified tensile strength of prestressing steel	270 ksi
I_g	moment of inertia of the gross concrete section	212,450 in^4
M_g	midspan moment due to member self-weight	632 ft-kips
M_D	midspan moment due to the slab dead load	600 ft-kips

Find the prestress loss due to elastic shortening, Δf_{pES}.

A Art. 5.9.5.2.3a-1

$$\Delta f_{pES} = \frac{E_p}{E_{ci}} f_{cgp}$$

Find the initial prestress before transfer, but after the changes due to the elastic deformations of the section, f_{pbt}, it is,

A Com. C5.9.5.2.3a

$$f_{pbt} = 0.9\ f_{pj}$$

$$f_{pj} = 0.75\ f_{pu} = (0.75)(270\ \text{ksi}) = 203\ \text{ksi}$$

A Tbl. 5.9.3-1

$$f_{pbt} = (0.9)(203\ \text{ksi}) = 182.3\ \text{ksi}$$

To avoid iteration, alternately the loss due to elastic shortening may be determined

A Eq. C5.9.5.2.3a-1

$$\Delta f_{pES} = \frac{A_{ps} f_{pbt}\left(I_g + e_m^2 A_g - e_m M_g A_g\right)}{A_{ps}\left(I_g + e_m^2 A_g\right) + \left(\dfrac{A_g I_g E_{ci}}{E_p}\right)}$$

$$= \frac{\begin{array}{c}\left(6.732\ \text{in}^2\right)\left(182.3\ \text{ksi}\right)\left(212{,}450\ \text{in}^4 + \left(18.38\ \text{in}\right)^2\left(762\ \text{in}^2\right)\right) \\ -\left(18.38\ \text{in}\right)\left(632\ \text{ft-kips}\right)\left(762\ \text{in}^2\right)\end{array}}{\begin{array}{c}\left(6.732\ \text{in}^2\right)\left(\left(212{,}450\ \text{in}^4\right) + \left(18.38\ \text{in}\right)^2\left(762\ \text{in}^2\right)\right) \\ + \left(\dfrac{\left(762\ \text{in}^2\right)\left(212{,}450\ \text{in}^4\right)\left(4696\ \text{ksi}\right)}{28{,}500\ \text{ksi}}\right)\end{array}}$$

$$= 15.77\ \text{ksi}$$

The reduced prestress force after transfer is,

$$P = (f_{pbt} - \Delta f_{pES})(A_{ps}) = (182.3\ \text{ksi} - 15.77\ \text{ksi})(6.732\ \text{in}^2)$$

$$= 1121.1\ \text{kips}$$

Concrete stress at top fiber is

$$f_t = \frac{-P}{A_g} + \frac{Pe_m}{S_{nc,top}} - \frac{M_g}{S_{nc,top}}$$

$$= \frac{-1121.1 \text{ kips}}{762 \text{ in}^2} + \frac{(1121.1 \text{ kips})(18.38 \text{ in})}{7692 \text{ in}^3} - \frac{(632 \text{ ft-kips})\left(12\frac{\text{in}}{\text{ft}}\right)}{7692 \text{ in}^3}$$

$$= -1.47 \text{ ksi} + 2.68 \text{ ksi} - 0.98 \text{ ksi}$$

$$= 0.23 \text{ ksi (tension)} < f_{ti} = 0.563 \text{ ksi [OK]}$$

Concrete stress in bottom fiber is

$$f_t = \frac{-P}{A_g} - \frac{Pe_m}{S_{nc,bottom}} + \frac{M_g}{S_{nc,bottom}}$$

$$= \frac{-1121.1 \text{ kips}}{762 \text{ in}^2} - \frac{(1121.1 \text{ kips})(18.38 \text{ in})}{9087 \text{ in}^3} + \frac{(632 \text{ ft-kips})\left(12\frac{\text{in}}{\text{ft}}\right)}{9087 \text{ in}^3}$$

$$= -1.47 \text{ ksi} - 2.27 \text{ ksi} + 0.83 \text{ ksi}$$

$$= -2.91 \text{ ksi (compression)} < f_{ci} = 3.6 \text{ ksi [OK]}$$

Please see Figure 2.68.

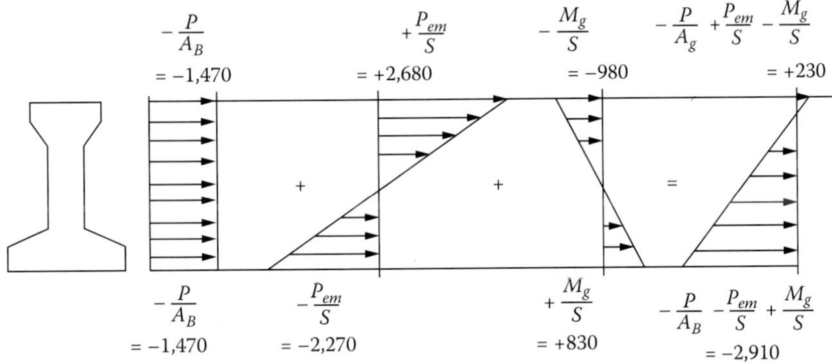

FIGURE 2.68
Concrete stresses at midspan at release of prestress for girder I-beam.

Step 5: Determine Final Concrete Stresses at Midspan after All Losses (Except Friction)

Find losses due to long-term shrinkage and creep of concrete, and low relaxation strand.

A Art. 5.4.2.3.1; Tbl. 5.9.3-1

Prestressing steel stress immediately prior to transfer

f_{pbt} = 0.75 f_{pu}, not including the elastic deformation. Thus, $f_{pi} = f_{pj}$ 203 ksi

H average annual ambient relative humidity 70% (for central NY)

A Fig. 5.4.2.3.3-1

Δf_{pR} an estimate of relaxation loss taken as 2.4 ksi for low relaxation strand, 10.0 ksi for stress relieved strand, and in accordance with manufacturer's recommendation for other types of strand.

A Art. 5.9.5.3

Determine time-dependent losses:

A Art. 5.9.5.3

The correction factor for relative humidity of the ambient air is

A Eq. 5.9.5.3-2

$$\gamma_h = 1.7 - 0.01\ H = 1.7 - (0.01)(70\%) = 1.0$$

The correction factor for specified concrete strength at time of prestress transfer to the concrete member is,

A Eq. 5.9.5.3-3

$$\gamma_{st} = \frac{5}{1 + f'_{ci}}$$

$$= \frac{5}{1 + 6.0\ \text{ksi}} = 0.833$$

The long-term prestress loss due to creep of concrete, shrinkage of concrete, and relaxation of steel is,

A Eq. 5.9.5.3-1

$$\Delta f_{pLT} = 10.0\frac{f_{pi}A_{ps}}{A_g}\gamma_h\gamma_{st} + 12.0\gamma_j\gamma_{st} + \Delta f_{pR}$$

$$= 10.0\left(\frac{(203\text{ ksi})(6.732\text{ in}^2)}{762\text{ in}^3}\right)(1.0)(0.833) + 12.0(1.0)(0.833) + 2.4\text{ ksi}$$

$$= 25.4\text{ ksi}$$

The total loss for pretensioned members is

A Eq. 5.9.5.1-1

$$\Delta f_{pT} = \Delta f_{pES} + \Delta f_{pLT}$$

$$= 15.77\text{ ksi} + 25.4\text{ ksi} = 41.17\text{ ksi}$$

The effective steel prestress after losses is

$$f_{se} = f_{pi}\ (= f_{pj}) - \Delta f_{pT} = 203\text{ ksi} - 41.7\text{ ksi} = 161.3\text{ ksi}$$

The effective prestress force after all losses, P_e, is

$$P_e = f_{se}A_{ps} = (161.3\text{ ksi})(6.732\text{ in}^2) = 1085.87\text{ kips } (1{,}085{,}870\text{ lbf})$$

Determine final concrete stresses at midspan after the total losses. For the basic beam section at the beam's bottom fiber,

$$\frac{-P_e}{A_g} - \frac{P_e e_m}{S_{nc,bottom}} + \frac{M_g + M_D}{S_{nc,bottom}} = \frac{-1{,}085{,}870\text{ lbf}}{762\text{ in}^2} - \frac{(1{,}085{,}870\text{ lbf})(18.38\text{ in})}{9087\text{ in}^3}$$

$$+\left(\frac{632\text{ ft-kips} + 600\text{ ft-kips}}{7692\text{ in}^3}\right)\left(12\frac{\text{in}}{\text{ft}}\right)\left(1000\frac{\text{lbf}}{\text{kip}}\right)$$

$$= -1994.4\text{ psi}$$

where:
M_g = moment due to beam weight
M_D = moment due to slab dead load

At the basic beam's top fiber, the final concrete stresses at midspan are

$$\frac{-P_e}{A_g} + \frac{P_e e_m}{S_{nc,top}} - \frac{M_g + M_D}{S_{nc,top}} = \frac{-1{,}085{,}870\text{ lbf}}{762\text{ in}^2} + \frac{(1{,}085{,}870\text{ lbf})(18.38\text{ in})}{7692\text{ in}^3}$$

$$-\left(\frac{632\text{ ft-kips} + 600\text{ ft-kips}}{7692\text{ in}^3}\right)\left(12\frac{\text{in}}{\text{ft}}\right)\left(1000\frac{\text{lbf}}{\text{kip}}\right)$$

$$= -752.3\text{ psi}$$

Moment due to the superimposed dead loads, M_s, consists of the superimposed dead load, w_s, the parapet/curb load (0.506 kip/ft), distributed equally to the 5 girders, plus the load of the 3 in future wearing surface,

$$w_s = \frac{\left(0.506\frac{\text{kip}}{\text{ft}}\right)(2)}{5 \text{ girders}} + (3 \text{ in})\left(\frac{\text{ft}}{12 \text{ in}}\right)\left(0.14\frac{\text{kips}}{\text{ft}^3}\right)(7.5 \text{ ft})$$

$$= 0.4649 \text{ kips/ft}$$

$$M_s = \frac{w_s L^2}{8} = \frac{\left(0.4649\frac{\text{kips}}{\text{ft}}\right)(80 \text{ ft})^2}{8}$$

$$= 371.9 \text{ ft-kips}$$

For composite section at the beam base,

$$\frac{M_s + M_{LL+IM}}{S_{bc}} = \left(\frac{371.9 \text{ ft-kips} + 1364.38 \text{ ft-kips}}{14{,}538.0 \text{ in}^3}\right)\left(12\frac{\text{in}}{\text{ft}}\right)\left(1000\frac{\text{lbf}}{\text{kip}}\right)$$

$$= 1433 \text{ psi}$$

At slab top,

$$-\frac{M_s + M_{LL+IM}}{S_{tc}} = \left(\frac{371.9 \text{ ft-kips} + 1364.38 \text{ ft-kips}}{24{,}476.6 \text{ in}^3}\right)\left(12\frac{\text{in}}{\text{ft}}\right)\left(1000\frac{\text{lbf}}{\text{kip}}\right)$$

$$= -851.2 \text{ psi}$$

See Figure 2.69.

Check the allowable concrete stresses at Service I Limit State load combination after all losses.

A Art. 5.9.4.2

For compression at girder top,

A Tbl. 5.9.4.2.1-1

$$f_{cs} = 0.45\, f'_c = (0.45)(6.5 \text{ ksi}) = 2.93 \text{ ksi} < -1311.1 \text{ ksi [OK]}$$

For tension at girder bottom,

A Tbl. 5.9.4.2.2-1

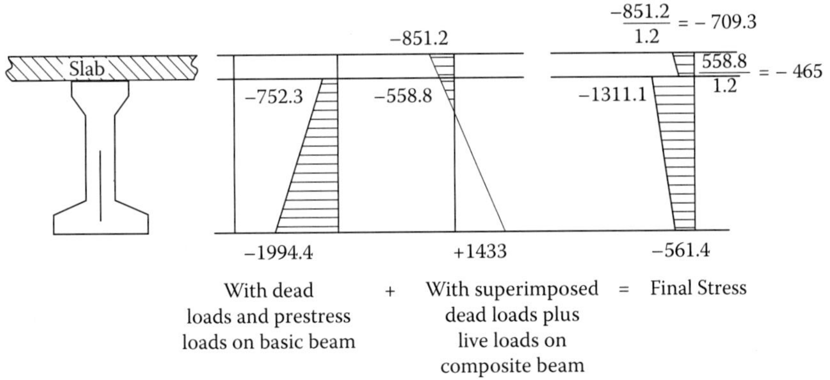

FIGURE 2.69
Final concrete stresses at midspan after losses.

$$f_{ts} = 0.19\sqrt{f_c'} = 0.19\sqrt{6.5 \text{ ksi}} = 0.484 \text{ ksi } > -0.5614 \text{ ksi [OK]}$$

Design Example 7: Flexural and Transverse Reinforcement for 50 ft Reinforced Concrete Girder

Situation

BW	barrier weight (curb, parapet, and sidewalk)	0.418 kips/ft
$f'_{c,beam}$	beam concrete strength	4.5 ksi
$f'_{c,deck}$	deck concrete strength	3.555 ksi
DW	future wearing surface load	2 in
L	bridge span	50 ft
LL	live load	HL-93
E_s	modulus of elasticity of reinforcing steel	29,000 ksi
f_y	specified minimum yield strength of reinforcing steel	60 ksi
S	girder spacing	8.0 ft
t_s	structural deck thickness	8.0 in
w_c	concrete unit weight	150 lbf/ft³

Please see Figure 2.70.

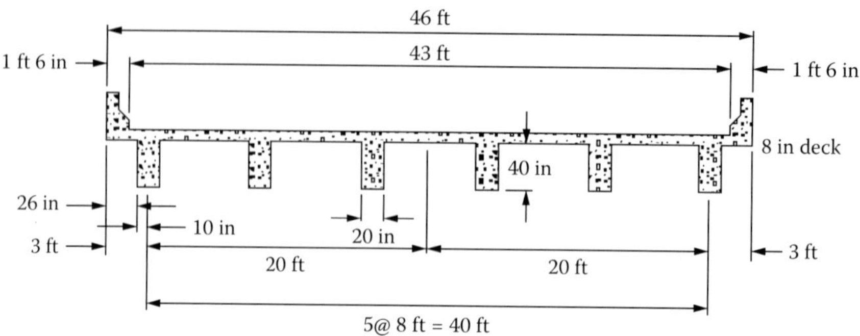

FIGURE 2.70
Reinforced concrete girder design example.

Requirements

Determine the flexural and transverse reinforcement for the reinforced concrete girder described.

Solution

The effective flange width of a concrete deck slab may be taken as the tributary width perpendicular to the axis of the member.

A Art. 4.6.2.6.1

$$b_e = 8 \text{ ft or } 96 \text{ in}$$

Step 1: Calculate the Beam Properties

The modulus of elasticity for the deck is

A Eq. 5.4.2.4.1

$$E_{deck} = 33{,}000\, w_c^{1.5}\sqrt{f_c'} = (33{,}000)\left(0.15\frac{\text{kips}}{\text{ft}^3}\right)^{1.5}\sqrt{3.555 \text{ ksi}}$$

$$= 3615 \text{ ksi}$$

The modulus of elasticity for the beam is

$$E_{deck} = 33{,}000\, w_c^{1.5}\sqrt{f_c'} = (33{,}000)\left(0.15\frac{\text{kips}}{\text{ft}^3}\right)^{1.5}\sqrt{4.5 \text{ ksi}}$$

$$= 4067 \text{ ksi}$$

The modular ratio is

A Eq. 4.6.2.2.1-2

$$n = \frac{E_{deck}}{E_{beam}} = \frac{3615 \text{ ksi}}{4067 \text{ ksi}} = 0.889$$

The transformed effective deck width is

$$b_{et} = (S)(n) = (8 \text{ ft})(0.889)(12 \text{ in/ft}) = 85.34 \text{ in}$$

The area of the T-beam is

$$A = (40 \text{ in})(20 \text{ in}) + b_{et}t_s = 800 \text{ in} + (85.34 \text{ in})(8 \text{ in}) = 1482.72 \text{ in}^2$$

$$= 10.30 \text{ ft}^2$$

From bottom, the center of gravity of T-beam $y_b = \Sigma Ay/\Sigma A$, is

$$y_b = \frac{(40 \text{ in})(20 \text{ in})(20 \text{ in}) + (85.34 \text{ in})(8 \text{ in})(40 \text{ in} + 4 \text{ in})}{(40 \text{ in})(20 \text{ in}) + (85.34 \text{ in})(8 \text{ in})}$$

$$= 31.05 \text{ in from bottom}$$

Moment of inertia of T-beam about the center of gravity is

$$I_g = I = \frac{bd^3}{12} + Ad^3$$

$$I_g = \left(\frac{1}{12}\right)(20 \text{ in})(40 \text{ in})^3 + (20 \text{ in})(40 \text{ in})(31.05 \text{ in} - 20 \text{ in})^2 = 322{,}430 \text{ in}^4$$

$$+ \frac{(85.34 \text{ in})(8 \text{ in})^3}{12} + (85.3 \text{ in})(8 \text{ in})(44 \text{ in} - 31.05 \text{ in})^2$$

$$= 15.55 \text{ ft}^4$$

Step 2: Calculate the Dead Load Effects

A Art. 3.3.2

The weight of the structural components is (see Figure 2.71)

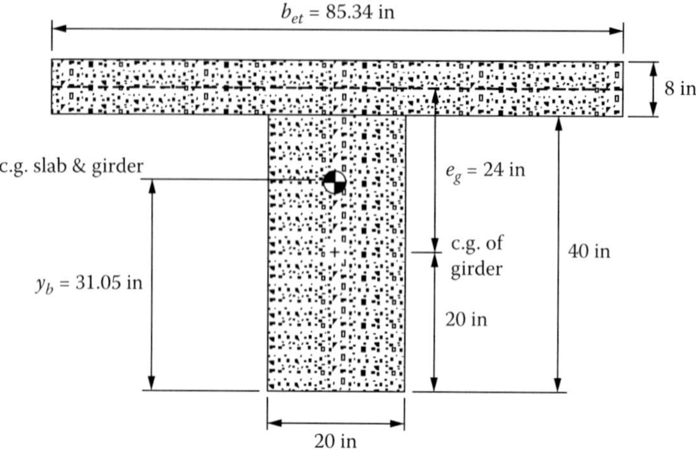

FIGURE 2.71
Girder section with area transformed deck slab.

$$DC_{T-beam} = (A)(w_c) = (10.30\ \text{ft}^2)\left(150 \frac{\text{lbf}}{\text{ft}^3}\right) = 1545 \frac{\text{lbf}}{\text{ft}}$$

Permanent loads such as barriers may be evenly distributed across all beams. The weight of the barriers is

$$w_{DC,barrier} = \frac{2(BW)}{\text{no. of beams}} = \frac{(2)\left(418 \frac{\text{lbf}}{\text{ft}}\right)}{6} = 139.33 \frac{\text{lbf}}{\text{ft}}$$

$$w_{DC,total} = w_{beam} + w_{barrier} = 1545\ \text{lbf/ft} + 139.33\ \text{lbf/ft}$$

$$= 1684.3\ \text{lbf/ft}$$

The weight of the wearing surface and utilities is

$$w_{DW} = \frac{(DW)w_cL}{\text{no. of beams}} = \frac{(2\ \text{in})\left(\frac{1\ \text{ft}}{12\ \text{in}}\right)\left(150 \frac{\text{lbf}}{\text{ft}}\right)(43\ \text{ft})}{6} = 179.2 \frac{\text{lbf}}{\text{ft}}$$

The unfactored maximum moments due to dead loads are

$$M_{DC} = \frac{w_{DC,total}L^2}{8} = \frac{\left(1684.3\,\frac{lbf}{ft}\right)(50\ ft)^2}{8}$$

$$= 5.26 \times 10^5\ ft\text{-}lbf$$

$$M_{DW} = \frac{w_{FWS}L^2}{8} = \frac{\left(179.2\,\frac{lbf}{ft}\right)(50\ ft)^2}{8}$$

$$= 0.56 \times 10^5\ ft\text{-}lbf$$

The unfactored maximum moments occur at midspan.
The unfactored maximum shears due to dead loads are

$$V_{DC} = \frac{w_{DC,total}L}{2} = \frac{\left(1684.3\,\frac{lbf}{ft}\right)(50\ ft)}{2}$$

$$= 42{,}107\ lbf$$

$$V_{DW} = \frac{w_{DW}L}{2} = \frac{\left(179.2\,\frac{lbf}{ft}\right)(50\ ft)}{2}$$

$$= 4{,}480\ lbf$$

The unfactored maximum shears occur at reactions for simply supported beams.

Step 3: Calculate the Live Load Effects

A Art. 3.6.1.2

The design live load, HL-93, shall consist of a combination of

- Design truck (HS-20) or design tandem and
- Design lane load

Dynamic load allowance, IM, is 33% for all limit states other than fatigue.

A Tbls. 3.6.2.1-1; 4.6.2.2.2b-1 or Appendix A

Calculate the live load distribution factor for moment, DFM, for interior beams of a typical cross section, e,

A Tbl. 4.6.2.2.1-1

The distance between the centers of gravity of the slab and the beam is

$$e_g = \frac{20\text{ in} + 4\text{ in}}{\frac{12\text{ in}}{\text{ft}}} = 2\text{ ft}$$

The stiffness parameter is

$$K_g = n(I + Ae_g^2) = (0.889)(15.55\text{ ft}^4 + (10.30\text{ ft}^2)(2\text{ ft})^2) = 50.44\text{ ft}^4$$

$$= 1.046 \times 10^6\text{ in}^4\ (10{,}000 \le K_g \le 7 \times 10^6\text{ in}^4)\ [\text{OK}]$$

A Eq. 4.6.2.2.1-1

Calculate the live load distribution factor for moment if the following are true.

A Tbl. 4.6.2.2.2b-1 or Appendix A

$$3.5 < S < 16\ [\text{OK}]$$

$$4.5 < t_s < 12\ [\text{OK}]$$

$$20 < L < 240\ [\text{OK}]$$

$$N_b > 4\ [\text{OK}]$$

The live load distribution factor for moment for interior beams with one design lane loaded is

$$DFM_{si} = 0.06 + \left(\frac{S}{14}\right)^{0.4}\left(\frac{S}{L}\right)^{0.3}\left(\frac{K_g}{12\,Lt_s^3}\right)^{0.1}$$

$$= 0.06 + \left(\frac{8\text{ ft}}{14}\right)^{0.4}\left(\frac{8\text{ ft}}{50\text{ ft}}\right)^{0.3}\left(\frac{1.05 \times 10^6\text{ in}^4}{\left(\frac{12\text{ in}}{\text{ft}}\right)(50\text{ ft})(8\text{ in})^3}\right)^{0.1}$$

$$= 0.581$$

The live load distribution factor for moment for interior beams with two or more design lanes loaded is

$$DFM_{mi} = 0.075 + \left(\frac{S}{9.5}\right)^{0.6}\left(\frac{S}{L}\right)^{0.2}\left(\frac{K_g}{12\,Lt_s^3}\right)^{0.1}$$

$$= 0.075 + \left(\frac{8\text{ ft}}{9.5}\right)^{0.6}\left(\frac{8\text{ ft}}{50\text{ ft}}\right)^{0.2}\left(\frac{1.05\times10^6\text{in}^4}{\left(\frac{12\text{ in}}{\text{ft}}\right)(50\text{ ft})(8\text{ in})^3}\right)^{0.1}$$

$$= 0.781$$

Calculate the live load distribution factor for moment for exterior beams with two or more design lanes loaded.

A Tbl. 4.6.2.2.2d-1 or Appendix B

The distance from the centerline of the exterior web of the exterior beam to the interior edge of the curb or traffic barrier, d_e, is

$$d_e = 18\text{ in or }1.5\text{ ft}$$

The correction factor for distribution is

$$e = 0.77 + \frac{d_e}{9.1\text{ ft}} = 0.77 + \frac{1.5\text{ ft}}{9.1\text{ ft}} = 0.935$$

Because $-1.0 < d_e < 5.5$, the live load distribution factor for moment for exterior beams with two or more design lanes loaded is

$$DFM_{me} = (e)(DFM_{mi}) = (0.935)(0.781) = 0.730$$

Calculate the live load distribution factor for moment for exterior beams with one design lane loaded using the lever rule. Please see Figure 2.72.

A Tbl. 4.6.2.2.2d-1; A Tbl. 3.6.1.1.2-1; C3.6.1.1.2

$$\Sigma M_{@hinge} = 0$$

$$0 = \frac{P}{2}(7.5\text{ ft}) + \frac{P}{2}(1.5\text{ ft}) - R(8\text{ ft})$$

$$R = 0.563\text{ P}$$

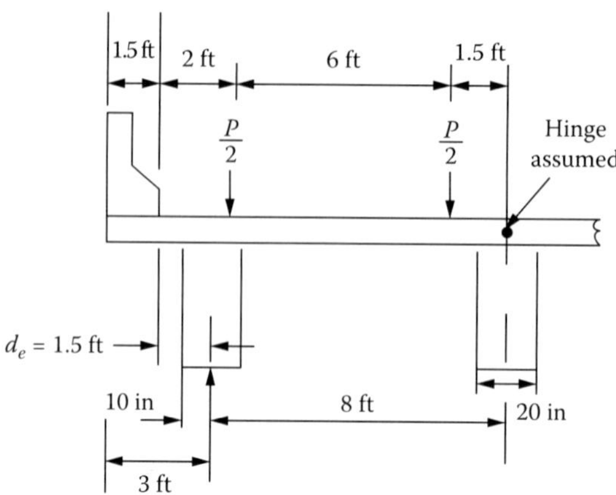

FIGURE 2.72
Lever rule for distribution factor for exterior girder moment with one lane loaded.

When the lever rule is used, the multiple presence factor, m, must be applied.

A Art. 3.6.1.1.2

The live load distribution factor for moment for exterior beams with one design lane loaded is

$$DFM_{se} = (m)(0.563) = (1.2)(0.563) = 0.676$$

$$DFM_{mi} = 0.781 \text{ for moment for interior beam [controls]}$$

The live load distribution factor for shear for interior beams with two or more design lanes loaded is

A Tbl. 4.6.2.2.3a-1 or Appendix C

$$DFV_{mi} = 0.2 + \frac{S}{12\text{ ft}} - \left(\frac{S}{35\text{ ft}}\right)^2 = 0.2 + \frac{8\text{ ft}}{12\text{ ft}} - \left(\frac{8\text{ ft}}{35\text{ ft}}\right)^2$$

$$= 0.814$$

The live load distribution factor for shear for interior beams with one design lane loaded is

$$DFV_{si} = 0.36 + \frac{S}{25\text{ ft}} = 0.36 + \frac{8\text{ ft}}{25\text{ ft}}$$

$$= 0.68$$

Calculate the live load distribution factor for shear for exterior beams with two or more design lanes loaded.

A Tbl. 4.6.2.2.3b-1 or Appendix D

The distance from the centerline of the exterior web of the exterior beam to the interior edge of the curb or traffic barrier, d_e, is 18 in or 1.5 ft.

The correction factor for distribution is

$$e = 0.6 + \frac{d_e}{10\text{ ft}} = 0.6 + \frac{1.5\text{ ft}}{10\text{ ft}} = 0.75$$

Because $-1.0 < d_e < 5.5$, the live load distribution factor for shear for exterior beams with two or more design lanes loaded is

$$DFV_{me} = (e)DFV_{int} = (0.75)(0.814) = 0.610$$

The live load distribution factor for shear for exterior beams with one design lane loaded using the lever rule is the same as for the moment DFM_{se}. Therefore,

$$DFV_{si} = 0.676$$

$$DFV_{mi} = 0.814 \text{ for shear for interior beams controls}$$

See Table 2.18.

Step 4: Calculate Truck Load (HS-20) and Lane Load Moments and Shears

A Art. 3.6.1.2

Calculate the unfactored flexural moment due to design truck load. (Refer to Design Example 1 and Figure 2.73.)

TABLE 2.18

Summary of Distribution Factors

Girder	DFM	DFV
Interior, two or more lanes loaded	0.781[a]	0.814[a]
Interior, one lane loaded	0.581	0.680
Exterior, two or more lanes loaded	0.730	0.610
Exterior, one lane loaded	0.676	0.676

[a] Controlling distribution factors.

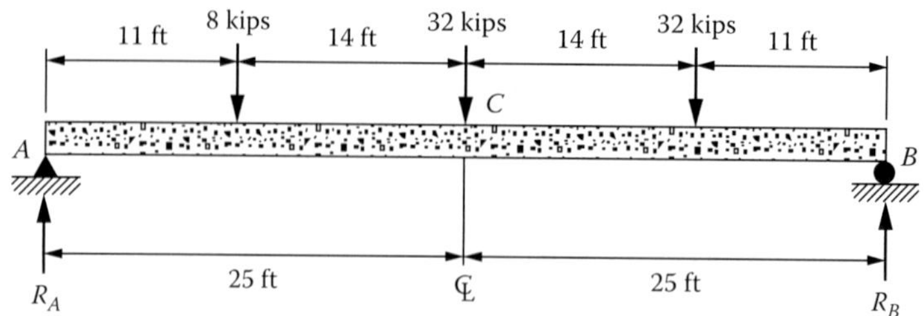

FIGURE 2.73
Design truck (HS-20) load position for the maximum moment at midspan.

$$\Sigma M_{@A} = 0$$

$$R_B = \frac{(8000\ \text{lbf})(11\ \text{ft}) + (32{,}000\ \text{lbf})(25\ \text{ft}) + (32{,}000\ \text{lbf})(39\ \text{ft})}{50\ \text{ft}}$$

$$= 42{,}720\ \text{lbf}$$

The unfactored flexural moment at midspan due to design truck load is

$$M_{tr} = (-32{,}000\ \text{lbf})(14\ \text{ft}) + (42{,}720\ \text{lbf})(25\ \text{ft})$$

$$= 6.2 \times 10^5\ \text{lbf-ft per lane}$$

(Also see Design Example 1.)

The unfactored flexural moment at midspan due to design tandem load is

$$M_{tandem} = 575{,}000\ \text{lbf-ft per lane}$$

(See Design Example 1)

Calculate the unfactored flexural moment due to lane load.

The ultimate moment due to live load of 640 lbf/ft occurs at the center of simply supported beam. See Figure 2.74.

A Art. 3.6.1.2.4

$$M_{ln} = \frac{w_{LL}L^2}{8} = \frac{\left(640\dfrac{\text{lbf}}{\text{ft}}\right)(50\ \text{ft})^2}{8}$$

$$= 2.0 \times 10^5\ \text{lbf-ft per lane}$$

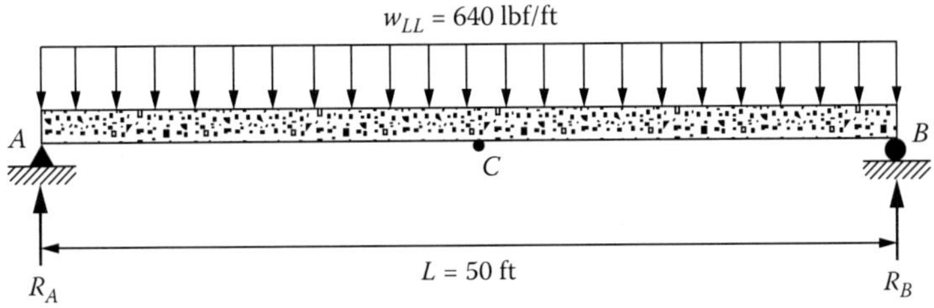

FIGURE 2.74
Design lane load moment at midspan.

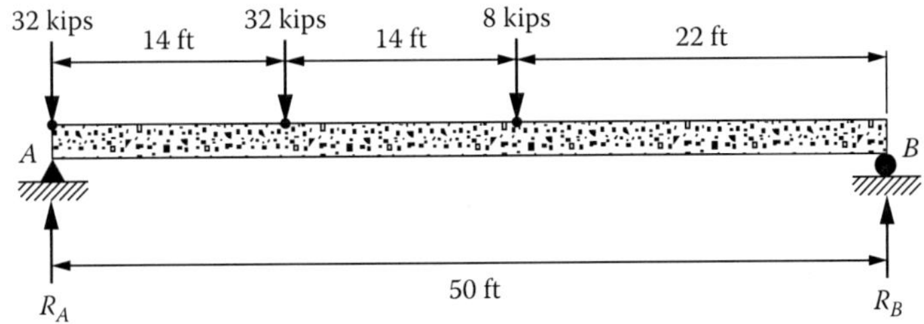

FIGURE 2.75
Design truck load (HS-20) load position for the maximum shear at support.

Design lane load is not subjected to a dynamic load allowance.

A Art. 3.6.1.2.4

The total unfactored flexural moment due to live loads per lane is (See Figure 2.75)

$$M_{LL+IM} = M_{tr}(1 + IM) + M_{ln} = (6.2 \times 10^5 \text{ lbf-ft})(1.33) + 2.0 \times 10^5 \text{ lbf-ft}$$

$$= 10.246 \times 10^5 \text{ lbf-ft per lane}$$

The total unfactored live load moment per girder is

$$M_{total} = M_{LL+IM}(DFM)$$

$$= (10.2 \times 10^5 \text{ lbf-ft})(0.781)$$

$$= 7.966 \times 10^5 \text{ lbf-ft per girder}$$

Calculate the unfactored shear due to design truck load.

$$\Sigma M_{@B} = 0$$

$$R_A = \frac{(8000\ \text{lbf})(22\ \text{ft}) + (32,000\ \text{lbf})(36\ \text{ft}) + (32,000\ \text{lbf})(50\ \text{ft})}{50\ \text{ft}}$$

$$= 58,560\ \text{lbf}$$

$$V = R_A = 58,560\ \text{lbf}$$

The unfactored shear due to design tandem load is

$$V_{tandem} = 48,000\ \text{lbf (See Design Example 1.)}$$

The unfactored shear due to design truck load is

$$V_{tr} = V(1 + IM) = (5.856 \times 10^4\ \text{lbf})(1.33)$$

$$= 7.79 \times 10^4\ \text{lbf}$$

Calculate the unfactored shear due to lane load.

The ultimate shear due to uniform live load of 640 lbf/ft occurring at the end of a simply supported beam is,

$$V_{lane} = R_A = \frac{w_{LL} L}{2} = \frac{\left(640\ \frac{\text{lbf}}{\text{ft}}\right)(50\ \text{ft})}{2}$$

$$= 1.6 \times 10^4\ \text{lbf per lane}$$

Dynamic effects are not taken into consideration for design lane loading. Please see Figure 2.76.

A Art. 3.6.1.2.4

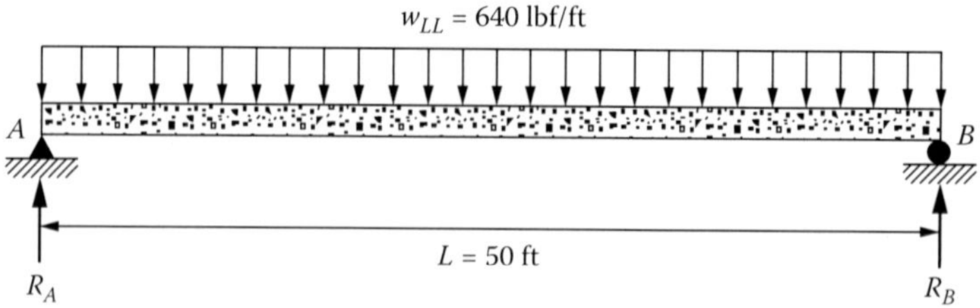

FIGURE 2.76
Design lane load position for the maximum shear at support.

The total unfactored shear per lane is

$$V_{LL+IM} = V_{tr} + V_{ln} = (7.79 \times 10^4 \text{ lbf}) + (1.6 \times 10^4 \text{ lbf})$$

$$= 9.39 \times 10^4 \text{ lbf per lane}$$

The total unfactored shear per interior girder is

$$V_{total} = V_{LL+IM}(DFV) = (9.39 \times 10^4 \text{ lbf})(0.814) = 7.64 \times 10^4 \text{ lbf}$$

Step 5: Calculate the Strength I Limit State

The factored moment for Strength I Limit State is

$$M_u = 1.25\, M_{DC} + 1.5\, M_{DW} + 1.75\, M_{total}$$

$$= (1.25)(5.26 \times 10^5 \text{ ft-lbf}) + (1.5)(0.56 \times 10^5 \text{ ft-lbf}) + (1.75)(7.996 \times 10^5 \text{ ft-lbf})$$

$$= 21.4 \times 10^5 \text{ ft-lbf}$$

The factored shear for Strength I Limit State is

$$V_u = 1.25\, V_{DC} + 1.5\, V_{DW} + 1.75\, V_{total}$$

$$= (1.25)(4.21 \times 10^4 \text{ lbf}) + (1.5)(0.448 \times 10^4 \text{ lbf}) + (1.75)(7.64 \times 10^4 \text{ lbf})$$

$$= 19.3 \times 10^4 \text{ lbf} = 193.0 \text{ kips}$$

Step 6: Design the Reinforcement

Design the flexural reinforcement. Neglect compression reinforcement in the flange. See Figure 2.77.

A Art. 5.14.1.5.1c; C5.14.1.5.1c

The web thickness with 5 #11 bars in each layer and #4 stirrup bars, b_w, is

$$b_w = (2)(1 \text{ in}) + (2)(0.5 \text{ in}) + 0.5\, d_b + (4)(1.5\, d_b)$$

$$= (2)(1 \text{ in}) + (2)(0.5 \text{ in}) + 0.5(1.4 \text{ in}) + (4)(1.5)(1.4 \text{ in})$$

$$= 18.4 \text{ in}$$

To allow for extra room, use $b_w = 20$ in.

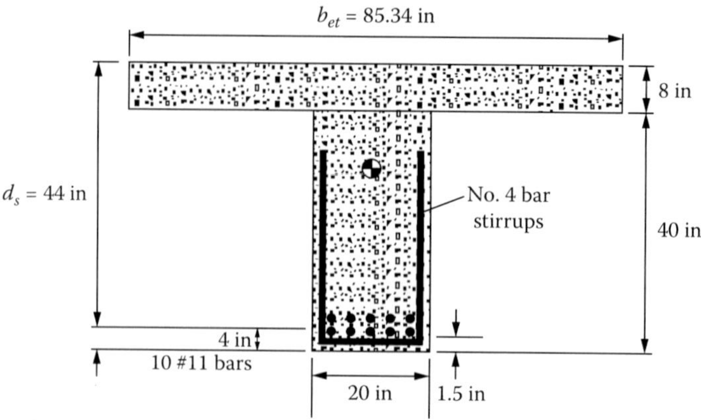

FIGURE 2.77
Reinforcement details.

Try using 10 no. 11 bars in two rows. The area of longitudinal reinforcement, A_s, is 15.6 in².

A Art. 5.10.3.1; 5.12.3

Assuming no. 4 stirrups and a 1.0 in concrete cover (assuming epoxy-coated bars), the distance from the girder base to the centroid of the reinforcement is

$$1.0\text{ in} + 0.5\text{ in} + d_b + \frac{d_b}{2} = 1.5\text{ in} + 1.41\text{ in} + \frac{1.41\text{ in}}{2} = 3.62\text{ in}\ \ (4\text{ in})$$

The stress block factor, β_1, is

A Art. 5.7.2.2; 5.7.3.1.1

$$\beta_1 = 0.85 - 0.05\left(f_c' - 4\text{ ksi}\right)$$

$$= 0.85 - 0.05(4.5\text{ ksi} - 4\text{ ksi})$$

$$= 0.825$$

The distance from the extreme compression fiber to the neutral axis is

$$c = \frac{A_s f_y}{0.85\, f_c'\, \beta_1 b_{et}} = \frac{\left(15.6\text{ in}^2\right)\left(60\text{ ksi}\right)}{\left(0.85\right)\left(4.5\text{ ksi}\right)\left(0.825\right)\left(85.34\text{ in}\right)} = 3.47\text{ in}$$

The depth of the equivalent rectangular stress block is

$$a = c\beta_1 = (3.47\text{ in})(0.825) = 2.86\text{ in}$$

The nominal resisting moment is

A Eq. 5.7.3.2.2-1

$$M_n = A_s f_y\left(d - \frac{a}{2}\right) = \left(15.6\text{ in}^2\right)\left(60\text{ ksi}\right)\left(44\text{ in} - \frac{2.86\text{ in}}{2}\right)\left(\frac{1\text{ ft}}{12\text{ in}}\right)$$

$$= 3320.5\text{ ft-kips}$$

The resistance factor, Φ, is 0.90.

A Art. 5.5.4.2.1

The factored resisting moment, M_r, is

A Eq. 5.7.3.2.1-1

$$M_r = \Phi M_n = (0.9)(3320.5\text{ ft-kips})$$

$$= 2988.5\text{ ft-kips} > M_u = 2140.0\text{ ft-kips [OK]}$$

Step 7: Check the Reinforcement Requirements

Review the flexural reinforcement.

The maximum flexural reinforcement provision was deleted in 2005.

A Art. 5.7.3.3.1

The minimum reinforcement requirement is satisfied if M_r (= ΦM_n) is at least equal to the lesser of

A Art. 5.7.3.3.2

- $M_r = \Phi M_n \geq 1.2\ M_{cr}$

or

- $M_r \geq$ (1.33)(the factored moment required by the applicable strength load combination)

where:

S_c	section modulus for the tensile extreme of the gross section	in^3
f_r	modulus of rupture of concrete, $0.37\sqrt{f'_c}$	ksi

A Art. 5.4.2.6

f'_c	girder concrete strength	4.5 ksi

y_b center of gravity of T-beam from tensile extreme face 31.05 in

I_g moment of inertia of the gross concrete section 322,430 in^4

$$M_{cr} = S_c f_r = \left(\frac{I_g}{y_b}\right) f_r = \left(\frac{322,430\ \text{in}^4}{31.05\ \text{in}}\right)\left(0.37\sqrt{4.5\ \text{ksi}}\right)$$

$$= 8150.45\ \text{kip-in} = 679.2\ \text{kip-ft}$$

A Eq. 5.7.3.3.2-1

$$1.2\ M_{cr} = (1.2)(679.2\ \text{kip-ft}) = 815.04\ \text{kip-ft, or}$$

$$1.33\ M_u = (1.33)\left(21.40 \times 10^5\ \text{lbf-ft}\right)\left(\frac{1\ \text{kip}}{1000\ \text{lbf}}\right)$$

$$= 2846.20\ \text{kip-ft}$$

$$1.2\ M_{cr} = 815.04\ \text{kip-ft [controls]}$$

$$M_r = \Phi M_n = 2988.5\ \text{kip-ft} > 1.2\ M_{cr} = 815.04\ \text{kip-ft [OK]}$$

The minimum reinforcement required is satisfied with 10 no. 11 bars as tension reinforcement.

Review the shear reinforcement.

A Art. 5.8.2.9

For simplicity, shear forces at end supports will be considered.

Find the factored shear for Strength I Limit State, V_s.

The effective shear depth need not be taken to be less than the greater of 0.9 d_e or 0.72 h.

$$d_v = d_s - \frac{a}{2} = 44\ \text{in} - \frac{2.86\ \text{in}}{2} = 42.57\ \text{in [controls]}$$

The minimum d_v is

$$d_v = 0.9\ d_e = (0.9)(44\ \text{in}) = 39.6\ \text{in}$$

or

$$d_v = 0.72\ h = (0.72)(48\ \text{in}) = 34.5\ \text{in}$$

See Figure 2.78.

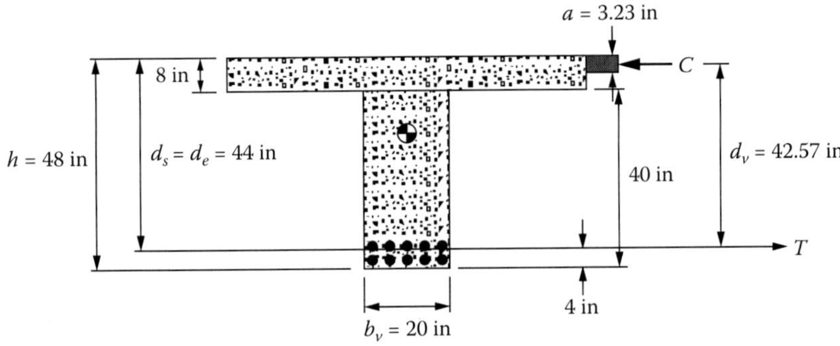

FIGURE 2.78
Review of shear reinforcement.

The transverse reinforcement shall be provided where

A Art. 5.8.2.4

$$V_u > 0.5\ \Phi V_c$$

A Eq. 5.8.2.4-1

β is 2.0 when Q = 45° for diagonal cracks.

A Art. 5.8.3.4.1; A Eq. 5.8.3.3-3

$$V_c = 0.0316\,\beta\sqrt{f'_c}\,b_v d_v = (0.0316)(2.0)\sqrt{4.5\text{ ksi}}\,(20\text{ in})(42.57\text{ in})$$

$$= 114.1\text{ kips}$$

$$0.5\ \Phi V_c = (0.5)(0.90)(114.1\text{ kips})$$

$$= 51.35\text{ kips}$$

A Eq. 5.8.2.4-1

$$V_u = 193\text{ kips} > 0.5\ \Phi V_c\ (= 51.35\text{ kips})$$

Therefore, the transverse reinforcement is needed at end support points.

A Art. 5.8.2.4

The nominal shear resistance, V_n, is the lesser of

A Art. 5.8.3.3

$$V_n = V_c + V_s$$

$$V_n = 0.25\ f'_c b_v d_v$$

V_s is the shear resistance provided by shear reinforcement.

$$V_n = 0.25\, f'_c b_v d_v = (0.25)(4.5\text{ ksi})(20\text{ in})(42.57\text{ in}) = 957.8\text{ kips}$$

Let the factored shear resistance, V_r, be equal to the factored shear load, V_u

A Eq. 5.8.2.1-2

$$V_r = \Phi V_n = V_u$$

$$V_n = \frac{V_u}{\Phi} = \frac{193.2\text{ kips}}{0.90} = 214.4\text{ kips [controls]}$$

The shear required by shear reinforcement, V_s, is found as follows.

A Eq. 5.8.3.3.1

$$V_n = V_c + V_s$$

$$V_s = V_n - V_c = \frac{V_u}{\Phi} - V_c = \frac{214.4\text{ kips}}{0.90} - 114.1\text{ kips}$$

$$= 124.1\text{ kips}$$

The angle of inclination of stirrups to longitudinal axis, α, is 90° and the angle of inclination of diagonal stress, θ, is 45°. Therefore, the shear resistance by shear reinforcement, V_s, and the spacing of stirrups, s, are calculated as follows.

Eq. C5.8.3.3-1, A Eq. 5.8.3.3.4; A Art. 5.8.3.4.1

$$V_s = \frac{A_v f_y d_v (\cot\theta + \cot\alpha)\sin\alpha}{s}$$

Using two #4 bar stirrups,

$$s = \frac{A_v f_y d_v \cot\theta}{V_s} = \frac{(2)(0.20\text{ in}^2)(60\text{ ksi})(42.57\text{ in})(1.0)}{124.1\text{ kips}}$$

$$= 8.2\text{ in [Try s = 8.0 in.]}$$

The minimum transverse reinforcement required is

A Eq. 5.8.2.5-1

$$A_v = 0.0316\sqrt{f'_c}\,\frac{b_v s}{f_y} = 0.0316\sqrt{4.5\text{ ksi}}\left(\frac{(20\text{ in})(8\text{ in})}{60\text{ ksi}}\right) = 0.18\text{ in}^2$$

The provided area of the shear reinforcement within the given distance, A_v, of two times 0.20 in^2 with no. 4 stirrups at 8 in spacing is OK.

Summary

For flexural reinforcement, use 10 no. 11 bars in 2 rows. For shear reinforcement, use no. 4 stirrups at 8 in spacing.

Design Example 8: Determination of Load Effects Due to Wind Loads, Braking Force, Temperature Changes, and Earthquake Loads Acting on an Abutment

A two-lane bridge is supported by seven W30 × 108 steel beams with 14 in × 0.6 in cover plates. The overall width of the bridge is 29.5 ft and the clear roadway width is 26.0 ft with an 8 in concrete slab. The superstructure span is 60.2 ft center to center on bearings (the overall beam length is 61.7 ft). It is 3.28 ft high including the concrete slab, steel beams, and 0.12 ft thick bearing.

The abutment has a total height of 16 ft and a length of 29.5 ft placed on a spread footing. The footing is placed on a gravel and sand soil.

Solution

Determine other load effects on the abutment.

1. Wind Loads

It is noted that the transverse wind loads (lateral to girder) in the plane of abutment length will be neglected because the abutment length is long (i.e., 29.5 ft).

Therefore, wind loads in longitudinal directions (perpendicular to the abutment wall face) will be considered.

1.A Wind Pressures on Structures, WS

A Art. 3.8.1.2

The skew angle as measured from a perpendicular to the beam longitudinal axis is 30°. See Figures 2.79 through 2.81.

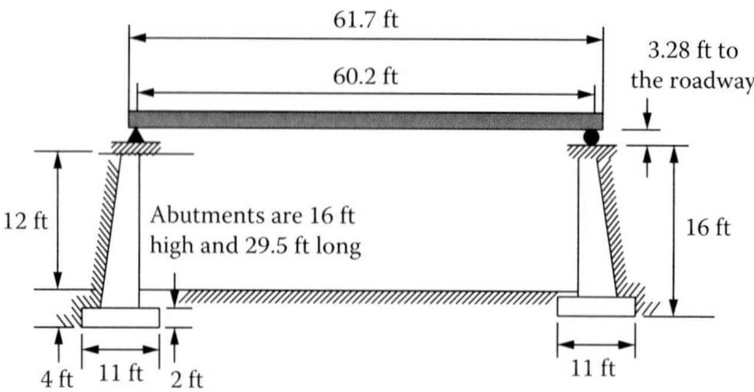

FIGURE 2.79
Abutment structure 16 ft in height and 29.5 ft in width.

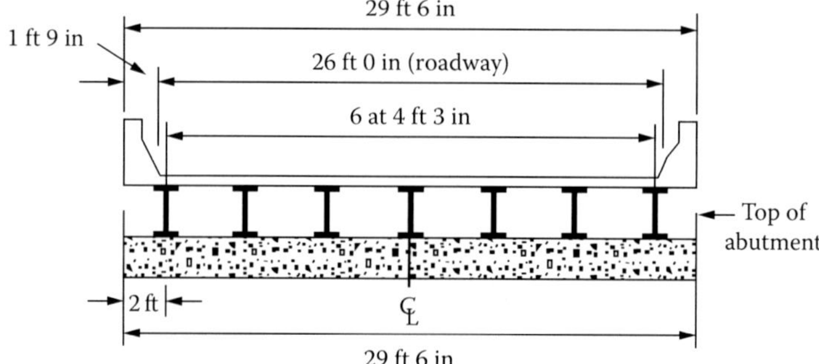

FIGURE 2.80
Two-lane bridge supported by seven W30 × 108 steel beams at 29.5 ft wide abutment.

Total depth away from abutment support = $h_{parapet} + t_{deck} + d_{girder} + t_{bearing}$,

$$= 32\text{ in}\left(\frac{1\text{ ft}}{12\text{ in}}\right) + 8\text{ in}\left(\frac{1\text{ ft}}{12\text{ in}}\right) + 2.49\text{ ft}(\text{Beam With No Cover Plate})$$

$$+0.05\text{ ft}(\text{Cover Plate})$$

$$= 5.87\text{ ft}$$

Total depth at abutment support = $h_{parapet} + t_{deck} + d_{girder} + t_{bearing}$,

$$= 32\text{ in}\left(\frac{1\text{ ft}}{12\text{ in}}\right) + 8\text{ in}\left(\frac{1\text{ ft}}{12\text{ in}}\right) + 2.49\text{ ft}(\text{Beam With No Cover Plate})$$

$$+0.12\text{ ft}(\text{Bearing Plate})$$

$$= 5.94\text{ ft}$$

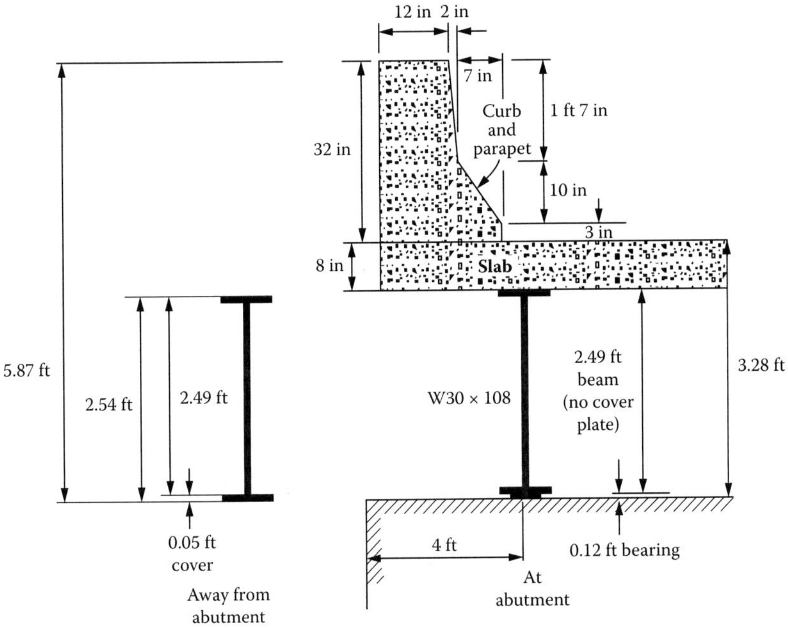

FIGURE 2.81
Steel beams at abutment and away from abutment.

$$\text{Overall beam length} = 61.70 \text{ ft}$$

Bearing to bearing length = 60.2 ft

A base design wind velocity at 30 ft height is,

A Art. 3.8.1.1

$$V_{30} = V_B = 100 \text{ mph}$$

Inasmuch as the abutment is less than 30 ft above ground level, the design wind velocity, V_{DZ}, does not have to be adjusted.

$$V_{DZ} = V_B$$

Design wind pressure, P_D, is

A Art. 3.8.1.2.1

$$P_D = P_B \left[\frac{V_{DZ}}{V_B} \right]^2$$

where:

P_B = base wind pressure specified in AASHTO [Tbl. 3.8.1.2.1-1] (kips/ft^2)

Base wind pressure, P_B, with skew angle of 30° is 0.012 kips/ft^2 for longitudinal load.

A Tbl. 3.8.1.2.2-1

The design longitudinal wind pressure, P_D, is

$$P_D = \left(0.012\frac{\text{kips}}{\text{ft}^2}\right)\left[\frac{100\ \text{mph}}{100\ \text{mph}}\right]^2$$

$$= 0.012\ \text{kips/ft}^2$$

The total longitudinal wind loading, WS_{total}, is

$$WS_{total} = (61.70\ \text{ft})(5.87\ \text{ft})(0.012\ \text{ksf})$$

$$WS_{total} = 4.35\ \text{kips}$$

The longitudinal wind force from super structure, WS_h, must be transmitted to the substructure through the bearings as follows:

4.35 kips wind loads from superstructure acts at its mid-depth (at the abutment) which is (5.87 ft – 0.05 ft cover plate)/2 + 0.12 ft bearing = 3.03 ft at the top of the abutment.

The longitudinal (horizontal) wind loading at the top of the abutment is

$$WS_h = 4.35\ \text{kips}/29.5\ \text{ft abutment length}$$

$$WS_h = 0.15\ \text{kips/ft abutment width}$$

The vertical wind loading at the top of the abutment, WS_v, is the reaction (see Figure 2.82)

$$(4.35\ \text{kips})(3.03\ \text{ft})/60.2\ \text{ft} = 0.22\ \text{kips up or down.}$$

Along the abutment width 29.5 ft, the vertical wind loading is,

$$WS_v = 0.22\ \text{kips}/29.5\ \text{ft}$$

$$= 0.007\ \text{kips/ft abutment width (negligible)}$$

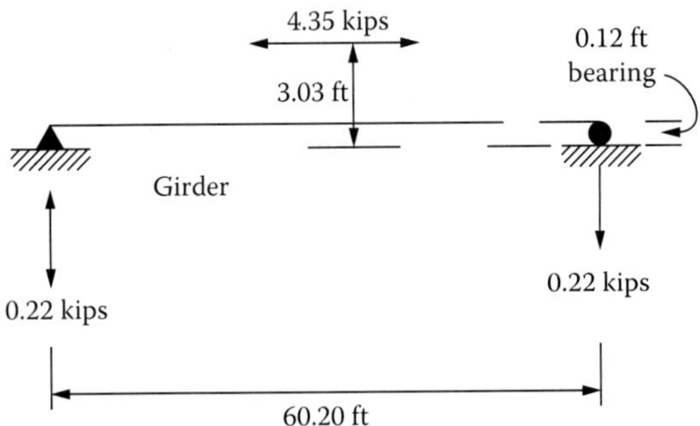

FIGURE 2.82
Wind loads on abutment transmitted from superstructure.

1.B Aeroelastic Instability

A Art. 3.8.3

$$\frac{\text{Span}}{\text{Abutment Width}} = \frac{60.2\text{ ft}}{29.5\text{ ft}} = 2.04$$

$$\frac{\text{Span}}{\text{Girder Depth}} = \frac{60.2\text{ ft}}{5.87\text{ ft}} = 10.2$$

The ratios are less than 30.0. Therefore, the bridge is deemed to be not wind sensitive.

1.C Wind Pressures on Vehicle Live Load, WL

A Art. 3.8.1.3, A Tbl. 3.8.1.3-1

$$\text{WL} = (0.024\text{ kips/ft for the skew angle of }30°)(61.7\text{ ft})$$

$$= 1.48\text{ kips acting at }6.0\text{ ft above the deck}$$

Or 1.48 kips will be acting at (6 ft + 3.28 ft) = 9.28 ft above the abutment top.

This 1.48 kips longitudinal force is transmitted to the abutment through the bearings as follows and as shown in Figure 2.83:

The reaction at the fixed abutment is

$$\frac{(1.48\text{ kips})(9.28\text{ ft})}{60.2\text{ ft}} = 0.228\text{ kips}$$

The longitudinal (horizontal) wind loading at the top of abutment due to vehicle live load is

$$\text{WL}_\text{h} = \frac{1.48\text{ kips}}{29.5\text{ ft}} = \frac{0.05\text{ kips}}{\text{ft abutment width}}$$

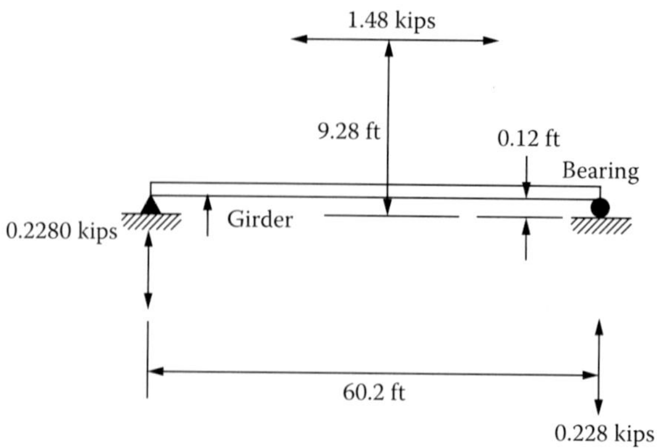

FIGURE 2.83
Wind loads on abutments transmitted from vehicle live load.

The vertical wind loading at the top of the abutment due to vehicle live load is

$$WL_v = \frac{0.228 \text{ kips}}{29.5 \text{ ft}} = \frac{0.0077 \text{ kips}}{\text{ft abutment width}} \text{ (negligible)}$$

1.D Vertical Wind Pressure

A Art. 3.8.2; Tbl. 3.4.1-1

A vertical upward wind force shall be applied only for Strength III and Service IV Limit States. Therefore, the vertical upward wind force is not applicable to Strength I Limit State.

1.E Wind Forces Applied Directly to the Substructure, W_{sub}

A Art. 3.8.1.2.3

The design wind pressure, P_D, is

$$P_D = P_B\left[\frac{V_{DZ}}{V_B}\right]^2$$

where:
P_B = base wind pressure
P_B = 0.04 kips/ft²

A Art. 3.8.1.2.3

$$P_D = P_B\left[\frac{V_{DZ}}{V_B}\right]^2$$

$$P_D = 0.04\frac{\text{kips}}{\text{ft}^2}\left[\frac{100\text{ mph}}{100\text{ mph}}\right]^2$$

$$P_D = 0.04\text{ kips/ft}^2$$

The wind loading of 0.04 kips/ft² will act, conservatively, perpendicular to the exposed 12 ft stem of the abutment, and it will act at 10 ft (= 12 ft/2 + 4 ft) above the base of the footing.

WS_{sub} = (0. 04 ksf)(12 ft) = 0.48 kips/ft abutment width

2. Braking Force, BR

A Art. 3.6.4

The braking force shall be taken as the greater of :

- 25% of the axle weights of the design truck or design tandem or
- 5% of the design truck plus lane load or
- 5% of the design tandem plus lane load

The static effects of the design truck or tandem for braking forces are not increased for dynamic load allowance.

A Art. 3.6.2.1

The total braking force is calculated based on the number of design lanes in the same direction.

A Art. 3.6.1.1; 3.6.4

In this example, it is assumed that all design lanes are likely to become one-directional in the future.

The braking force for a single traffic lane is as follows:

BR_{truck} = 25% of design truck = (0.25)(32 kips + 32 kips + 8 kips)

= 18.0 kips [controls]

BR_{tandem} = 25% of the design tandem = (0.25)(25 kips + 25 kips)

= 12.50 kips

$BR_{truck+lane}$ = 5% of the design truck plus lane load

= 0.05[(32 kips + 32 kips + 8 kips) + (0.64 kips/ft x 60.2 ft)]

= 7.45 kips

$BR_{tandem+lane}$ = 5% of the design tandem plus lane load

= 0.05[(25 kips + 25 kips) + (0.64 kips/ft x 60.2 ft)]

= 4.43 kips

The braking forces act horizontally at a distance 6 ft above the roadway surface in either longitudinal direction. The multiple presence factor in AASHTO [3.6.1.1.2] shall apply.

A Art. 3.6.4

For the number of loaded lanes of 2, multiple presence factor, m, is 1.0.

A Tbl. 3.6.1.1.2-1

Maximum braking force,

$$BR_{max} = BR_{truck} = 18.0 \text{ kips}$$

The longitudinal (horizontal) braking force is transmitted to the abutment as follows.

The longitudinal (horizontal) force at the top of the abutment due to braking force is (see also Figure 2.84):

$$BR_{hor} = \frac{18.0 \text{ kips}}{29.5 \text{ ft}} = \frac{0.61 \text{ kips}}{\text{ft abutment width}}$$

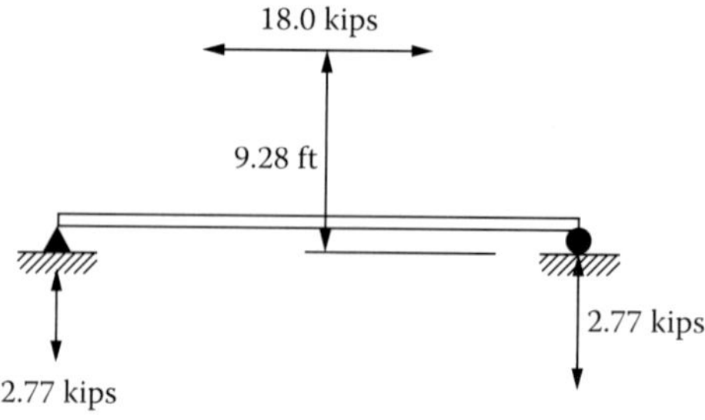

FIGURE 2.84
Forces on abutment from braking.

The vertical reaction at the abutment is:

$$\frac{18.0\text{ kips}\times 9.28\text{ ft}\left(=6\text{ ft above the roadway}+3.28\text{ ft}\right)}{60.2\text{ ft}}=2.77\text{ kips}$$

The vertical force at the top of the abutment due to braking force is:

$$BR_{vert}=\frac{2.77\text{ kips}}{29.5\text{ ft}}=\frac{0.094\text{ kips}}{\text{ft abutment width}}$$

3. Force Effect Due to Uniform Temperature Change

A Art. 3.12.2

Assume a moderate climate. The temperature range is 0°F to 120°F. Also, assume steel girder setting temperature T_{set} is 68°F.

A Tbl. 3.12. 2. 1-1

The thermal coefficient of steel is 6.5×10^{-6} in/in/°F.

A Art. 6.4.1

Expansion thermal movement

$$\Delta_{exp}=\left(6.5\times 10^{-6}\frac{\frac{\text{in}}{\text{in}}}{^\circ\text{F}}\right)\left(120^\circ\text{F}-68^\circ\text{F}\right)\left(60.2\text{ ft}\left(\frac{12\text{ in}}{1\text{ ft}}\right)\right)$$

$$=0.244\text{ in}$$

Contraction thermal movement

$$\Delta_{contr}=\left(6.5\times 10^{-6}\frac{\frac{\text{in}}{\text{in}}}{^\circ\text{F}}\right)\left(68^\circ\text{F}-0^\circ\text{F}\right)\left(60.2\text{ ft}\left(\frac{12\text{ in}}{1\text{ ft}}\right)\right)$$

$$=0.32\text{ in}$$

The elastomeric bearing pad properties assumed are:

Shear modulus of the elastomer, G, is

A Art. 14.7.5.2

$$G = 0.095 \text{ ksi}$$

$$0.08 \text{ ksi} < 0.095 \text{ ksi} < 0.175 \text{ ksi [OK]}$$

Pad area, A, is

A Art. 14.6.3.1; 14.7.2.3.1

$$A = 210.0 \text{ in}^2$$

Pad thickness = 3.5 in

Lateral (horizontal) loads due to temperature,

A Eq. 14.6.3.1-2

$$H_{bu} = GA\left(\frac{\Delta_u}{h_{rt}}\right)$$

where:
Δ_u = shear deformation
h_{rt} = elastomer thickness

The load due to expansion is

$$H_{rise} = (0.095 \text{ ksi})(210.0 \text{ in}^2)\left(\frac{0.244 \text{ in}}{3.5 \text{ in}}\right)$$

$$= 1.39 \text{ kips per bearing}$$

Multiply H_{rise} by seven beam bearings and divide by the abutment length to determine the lateral load at the top of the abutment.

$$H_{temp,rise} = \frac{(1.39 \text{ kips} \times 7 \text{ bearings})}{29.5 \text{ ft}} = 0.33 \frac{\text{kips}}{\text{ft abutment width}}$$

The load due to contraction is

$$H_{temp,fall} = (0.095 \text{ ksi})(210.0 \text{ in}^2)\left(\frac{0.32 \text{ in}}{3.5 \text{ in}}\right)$$

$$= 1.82 \text{ kips per bearing}$$

The lateral (horizontal) load at the top of the abutment due to temperature fall is calculated in the same manner as for temperature rise.

$$H_{temp\ fall} = \frac{(1.82\ \text{kips} \times 7\ \text{bearings})}{29.5\ \text{ft}} = 0.43 \frac{\text{kips}}{\text{ft abutment width}} \left[\text{controls}\right]$$

4. Earthquake Loads

A Art. 3.10; 3.10.9; 4.7.4.1; 4.7.4.2

First Step: Describe the type of bridge, number of spans, height of piers, type of foundations, subsurface soil conditions, and so on.

Second Step: Determine the horizontal acceleration coefficient, A, that is appropriate for bridge site (Art. 3.10.2; Fig. 3.10.2.1-1 and 3.10.2.1-2; Tbl. 3.10.3.1-1) and seismic zones (Art. 3.10.9; Art. 3.10.6). The values of the coefficients in the contour maps are expressed in percent. Thus, numerical values for the coefficient are obtained by dividing the contour values by 100. Local maxima and minima are given inside the highest and lowest contour for a particular region.

A Art. 3.10.2

Third Step: Determine the seismic performance zone for the bridge.

A Art. 3.10.6

Fourth Step: Determine the importance category of the bridge.

A Art. 3.10.5

Fifth Step: Determine the bridge site coefficient, which is based on soil profile types defined in Art. 3.10.3.1 through 3.10.7.1.

A Art. 3.10.3

Sixth Step: Determine the response modification factors (R factors), which reduce the seismic force based on elastic analysis of the bridge. The force effects from an elastic analysis are to be divided by the R factors.

A Art. 3.10.7; Tbl. 3.10.7.1-1; 3.10.7.1-2

Based on the information from the six steps, the level of seismic analysis required can be determined.

Single-Span Bridges

A Art. 4.7.4.2

Seismic analysis is not required for single-span bridges, regardless of seismic zone.

Connections between the bridge superstructure and the abutment shall be designed for the minimum force requirements as specified in AASHTO

[3.10.9]. Minimum seat width requirements shall be satisfied at each abutment as specified in AASHTO [4.7.4.4].

Seismic Zone 1

A Art. 3.10.9.2, C3.10.9.2, 3.10.9

For New York City, the peak ground acceleration coefficient on rock (site class B), PGA, is 0.09.

A Fig. 3.10.2.1-1

Site Class B (Rock)

A Tbl. 3.10.3.1-1

Value of site factor, f_{pga}, for the site class B and PGA less than 0.10 is 1.0.

A Tbl. 3.10.3.2-1

The peak seismic ground acceleration, A_s, is

A Eq. 3.10.4.2-2

$$A_s = (f_{pga})(PGA) = (1.0)(0.09)$$

$$= 0.09 \geq 0.05$$

The horizontal design connection force in the restrained directions shall not be less than 0.25 times the vertical reaction due to the tributary permanent load and the tributary live loads assumed to exist during an earthquake.

A Art. 3.10.9.2

The magnitude of live load assumed to exist at the time of the earthquake should be consistent with the value of γ_{EQ} for the extreme event I Limit State used in conjunction with Table 3.4.1-1. γ_{EQ} issue is not resolved and therefore used as 0.0.

A Comm. 3.10.9.2

Because all abutment bearings are restrained in the transverse direction, the tributary permanent load can be taken as the reaction at the bearing. Therefore, no tributary live loads will be considered.

A Art. 3.10.9.2

The girder vertical reaction, DL = 78.4 kips [previously known].

The maximum transverse horizontal earthquake load = 0.25 x 78.4 kips = 19.6 kips.

TABLE 2.19

Summary of Forces

Component	Description	Forces per ft of Abutment Width (kips)
WS_h	Longitudinal (horizontal) wind loading	0.15
WS_v	Vertical wind loading	0.007
WL_h	Longitudinal (horizontal) wind loading due to vehicle live load	0.05
WL_v	Vertical wind loading due to vehicle live load	0.0077
WS_{sub}	Wind loading perpendicular to exposed abutment stem	0.48
BR_{hor}	Longitudinal (horizontal) force due to braking force	0.61
BR_{vert}	Vertical force due to braking force	0.094
$H_{temp,\ rise}$	Lateral load due to temperature rise	0.33
$H_{temp,\ fall}$	Lateral load due to temperature fall	0.43
EQ_h	Transverse (lateral) earthquake loading	0.66

The transverse (horizontal) earthquake loading at the top of the abutment is

$$EQ_h = \frac{19.6 \text{ kips}}{29.5 \text{ ft abutment}} = \frac{0.66 \text{ kips}}{\text{ft abutment width}}$$

Please see Table 2.19 and Figure 2.85 for the summary of the forces. For the load combinations and load factors, refer to AASHTO Table 3.4.1-1 for the appropriate limit states.

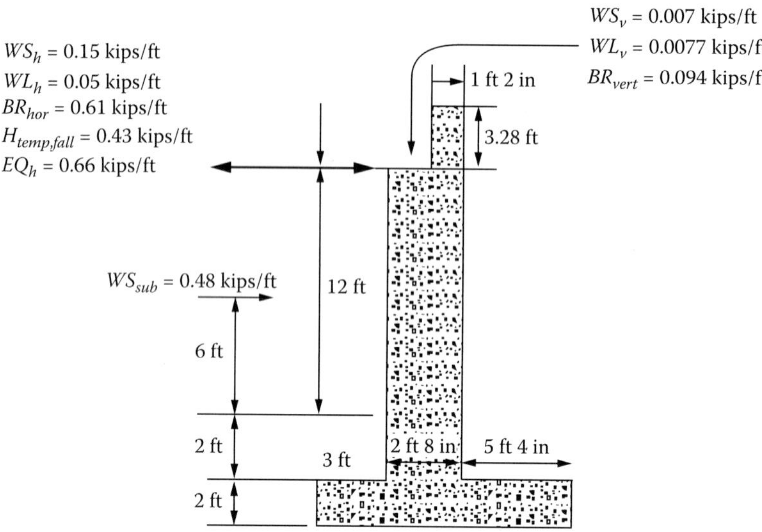

FIGURE 2.85
Summary of forces on abutment due to wind loads, braking forces, temperature changes, and earthquake loads.

3

Practice Problems

Practice Problem 1: Noncomposite 60 ft Steel Beam Bridge for Limit States Strength I, Fatigue II, and Service

Situation

L	bridge span	60 ft
LL	live load	HL-93
FWS	future wearing surface (bituminous concrete)	3 in
f'_c	concrete strength	5.0 ksi
ADTT	average daily truck traffic	250
	fatigue design life	75 years
	New Jersey barrier	0.45 kips/ft each
w	roadway width	43 ft
w_c	dead load of concrete	150 lbf/ft^3
w_w	dead load of wearing surface (bituminous concrete)	140 lbf/ft^3

Please see Figure 3.1.

Requirements

Review:

Flexural and shear resistance for Strength I Limit State

Fatigue II Limit State

Service Limit State

Elastic deflection

Flexure requirement to prevent permanent deflections

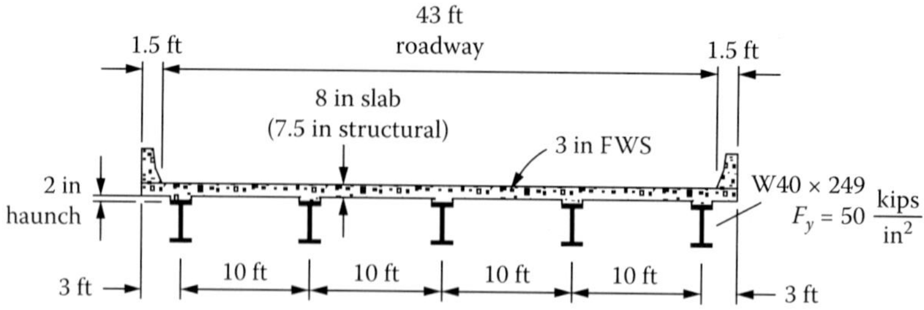

FIGURE 3.1
Cross section of noncomposite steel beam bridge.

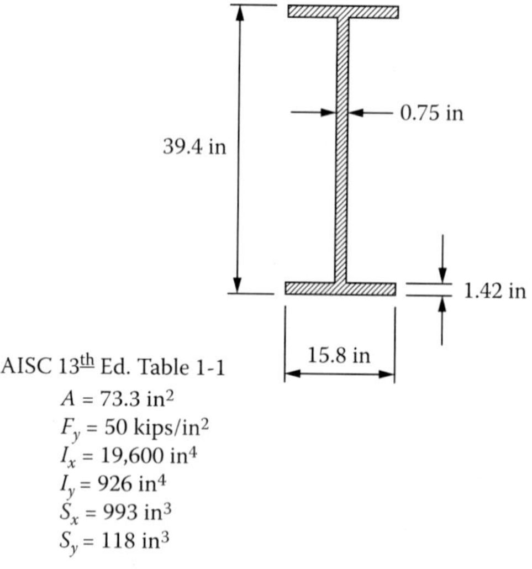

FIGURE 3.2
W40 × 249 properties.

Solution

Step 1: Review Section, W40 × 249

AISC (13th ed.) Tbl. 1-1

Please see Figure 3.2.

Deck slab, 8 in (7.5 in structural)

Effective Flange Width of a concrete deck slab be taken as the tributary width perpendicular to the axis of the member.

A Art. 4.6.2.6.1

t_s	slab thickness	8 in
b_i	effective flange width for interior beams	in
b_e	effective flange width for exterior beams	in
b_w	web width	0.75 in
b_f	flange width of steel section	15.8 in
S	average spacing of adjacent beams	10 ft

The effective flange width for interior beams is

$$b_i = S = 10 \text{ ft} = 120 \text{ in}$$

The effective flange width for exterior beams is

$$b_e = (3 \text{ ft})\left(\frac{12 \text{ in}}{\text{ft}}\right) + \left(\frac{10 \text{ ft}}{2}\right)\left(\frac{12 \text{ in}}{\text{ft}}\right) = 96 \text{ in}$$

Step 2: Evaluate Dead Loads

The structural components dead load, DC_1, per interior girder is calculated as follows and as shown in Figure 3.3.

$$DC_{slab} = \left(\frac{120 \text{ in}}{12\frac{\text{in}}{\text{ft}}}\right)\left(\frac{8 \text{ in}}{12\frac{\text{in}}{\text{ft}}}\right)\left(0.150\frac{\text{kips}}{\text{ft}^3}\right) = 1.0 \text{ kips/ft}$$

$$DC_{haunch} = \left(\frac{2 \text{ in}}{12\frac{\text{in}}{\text{ft}}}\right)\left(\frac{15.8 \text{ in}}{12\frac{\text{in}}{\text{ft}}}\right)\left(0.150\frac{\text{kips}}{\text{ft}^3}\right) = 0.033 \text{ kips/ft}$$

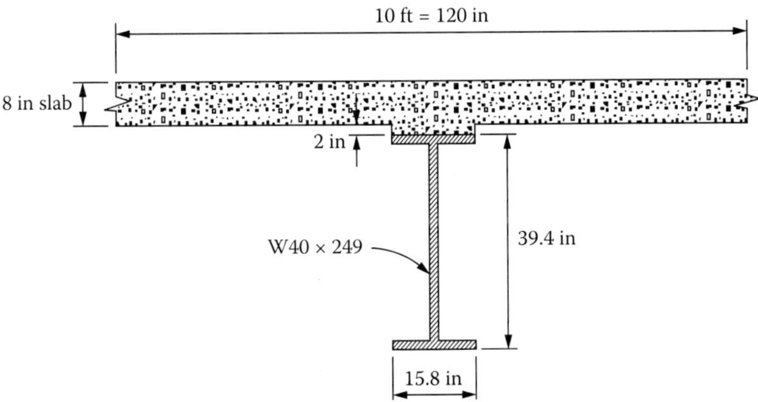

FIGURE 3.3
Dead loads for interior girder.

Assuming 5% of steel weight for diaphragms, stiffeners, and so on,

$$DC_{slab} = 0.249\frac{\text{kips}}{\text{ft}} + (0.05)\left(0.249\frac{\text{kips}}{\text{ft}}\right) = 0.261\ \text{kips/ft}$$

$$DC_{\text{stay-in-place forms}} = \left(7\frac{\text{lbf}}{\text{ft}^2}\right)\left(\frac{\text{roadway width}}{\text{no. of girders}}\right) = \left(0.007\frac{\text{kips}}{\text{ft}^2}\right)\left(\frac{43\ \text{ft}}{5}\right)$$

$$= 0.06\ \text{kips/ft}$$

The structural components dead load is

$$DC_1 = DC_{slab} + DC_{haunch} + DC_{steel} + DC_{\text{SIP forms}}$$

$$= 1.0\ \text{kips/ft} + 0.033\ \text{kip/ft} + 0.261\ \text{kips/ft} + 0.06\ \text{kips/ft} = 1.354\ \text{kips/ft}$$

The shear for DC_1 is

$$V_{DC_1} = \frac{wL}{2} = \frac{\left(1.354\frac{\text{kips}}{\text{ft}}\right)(60\ \text{ft})}{2} = 40.62\ \text{kips}$$

The moment for DC_1 is

$$M_{DC_1} = \frac{wL^2}{8} = \frac{\left(1.354\frac{\text{kips}}{\text{ft}}\right)(60\ \text{ft})^2}{8} = 609.3\ \text{ft-kips}$$

The nonstructural dead load, DC_2, per interior girder is calculated as

$$DC_{barrier} = \frac{\left(0.45\frac{\text{kips}}{\text{ft}}\right)(2\ \text{barriers})}{5\ \text{girders}} = 0.18\ \text{kips/ft}$$

The shear for DC_2 is

$$V_{DC_2} = \frac{wL}{2} = \frac{\left(0.18\frac{\text{kips}}{\text{ft}}\right)(60\ \text{ft})}{2} = 5.4\ \text{kips}$$

The moment for DC_2 is

$$M_{DC_2} = \frac{wL^2}{8} = \frac{\left(0.18\,\frac{\text{kips}}{\text{ft}}\right)(60\text{ ft})^2}{8} = 81.0\text{ ft-kips}$$

Total structural and nonstructural dead load, DC_{tot}, for interior girders is

$$V_{DC,tot} = V_{DC1} + V_{DC2} = 40.62\text{ kips} + 5.4\text{ kips} = 46.02\ (46\text{ kips})$$

$$M_{DC,tot} = M_{DC1} + M_{DC2} = 609.3\text{ ft-kips} + 81.0\text{ ft-kips} = 690.3\text{ ft-kips}$$

The wearing surface dead load, DW, for interior girders is

$$DW_{FWS} = \frac{(FWS)w_wL}{\text{no. of beams}} = \frac{(3\text{ in})\left(\frac{1\text{ ft}}{12\text{ in}}\right)\left(140\,\frac{\text{lbf}}{\text{ft}^3}\right)(43\text{ ft})}{5\text{ beams}}$$
$$= 300\text{ lbf/ft} = 0.3\text{ kips/ft}$$

The shear for DW is

$$V_{DW} = \frac{wL}{2} = \frac{\left(0.3\,\frac{\text{kips}}{\text{ft}}\right)(60\text{ ft})}{2} = 9.0\text{ kips}$$

The moment for DW is

$$M_{DW} = \frac{wL^2}{8} = \frac{\left(0.3\,\frac{\text{kips}}{\text{ft}}\right)(60\text{ ft})^2}{8} = 135\text{ ft-kips}$$

Find the structural components dead load, DC_1, per exterior girder (see Figure 3.4).

$$DC_{slab} = (8\text{ ft})\left(\frac{8\text{ in}}{12\,\frac{\text{in}}{\text{ft}}}\right)\left(0.150\,\frac{\text{kips}}{\text{ft}^3}\right) = 0.80\text{ kips/ft}$$

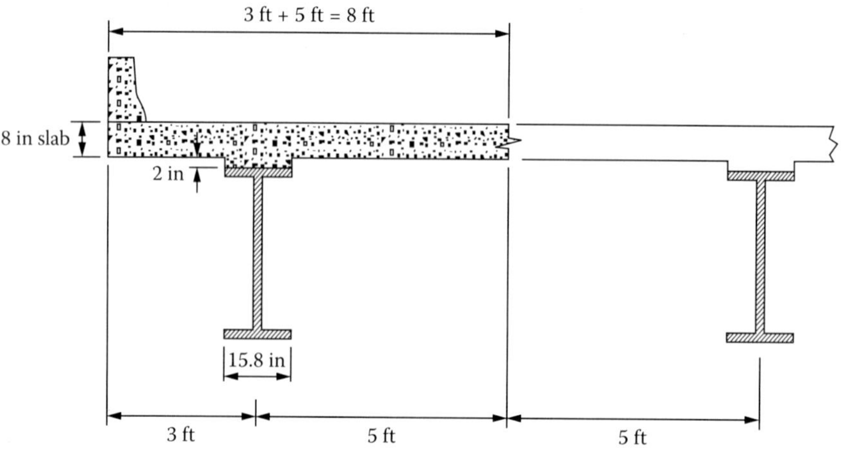

FIGURE 3.4
Dead loads for exterior girder.

$$DC_{haunch} = \left(\frac{2\text{ in}}{12\dfrac{\text{in}}{\text{ft}}}\right)(15.8\text{ in})\left(\frac{1\text{ ft}}{12\text{ in}}\right)\left(0.150\frac{\text{kips}}{\text{ft}^3}\right) = 0.033\text{ kips/ft}$$

$$DC_{steel} = \left(0.249\frac{\text{kips}}{\text{ft}}\right) + (0.05)\left(0.249\frac{\text{kips}}{\text{ft}}\right) = 0.261\text{ kips/ft}$$

$$DC_{stay\text{-}in\text{-}place\ forms} = \left(7\frac{\text{lbf}}{\text{ft}^2}\right)\left(\frac{\text{roadway width}}{\text{no. of girders}}\right)\left(0.007\frac{\text{kips}}{\text{ft}^2}\right)(43\text{ ft})\left(\frac{1}{5}\right)$$
$$= 0.06\text{ kips/ft}$$

The structural components dead load, DC_1, per exterior girder is

$$DC_1 = DC_{slab} + DC_{haunch} + DC_{steel} + DC_{SIP\ forms}$$
$$= 0.80\text{ kips/ft} + 0.033\text{ kip/ft} + 0.261\text{ kips/ft} + 0.06\text{ kips/ft} = 1.154\text{ kips/ft}$$

The shear for DC_1, per exterior girder is

$$V_{DC_1} = \frac{wL}{2} = \frac{\left(1.154\dfrac{\text{kips}}{\text{ft}}\right)(60\text{ ft})}{2} = 34.62\text{ kips}$$

The moment for DC_1, per exterior girder is

$$M_{DC_1} = \frac{wL^2}{8} = \frac{\left(1.154\frac{\text{kips}}{\text{ft}}\right)(60\text{ ft})^2}{8} = 519.3\text{ ft-kips}$$

The nonstructural dead load, DC_2, per exterior girder is

$$DC_{\text{barrier}} = \frac{\left(0.45\frac{\text{kips}}{\text{ft}}\right)(2\text{ barriers})}{5\text{ girders}} = 0.18\text{ kips/ft}$$

The shear for DC_2, per exterior girder is

$$V_{DC_2} = \frac{wL}{2} = \frac{\left(0.18\frac{\text{kips}}{\text{ft}}\right)(60\text{ ft})}{2} = 5.4\text{ kips}$$

The moment for DC_2, per exterior girder is

$$M_{DC_2} = \frac{wL^2}{8} = \frac{\left(0.18\frac{\text{kips}}{\text{ft}}\right)(60\text{ ft})^2}{8} = 81\text{ ft-kips}$$

The shear for the structural and nonstructural dead load, DC_{tot}, for exterior girder is

$$V_{DC,tot} = V_{DC1} + V_{DC2} = 34.62\text{ kips} + 5.4\text{ kips} = 40.0\text{ kips}$$

The moment for the structural and nonstructural dead load for exterior girder is

$$M_{DC,tot} = M_{DC1} + M_{DC2} = 519.3\text{ ft-kips} + 81\text{ ft-kips} = 600.3\text{ ft-kips}$$

The shear for the wearing surface dead load, DW, for exterior girders is

$$V_{DW} = \frac{wL}{2} = \frac{\left(0.3\frac{\text{kips}}{\text{ft}}\right)(60\text{ ft})}{2} = 9.0\text{ kips}$$

The moment for the wearing surface dead load, DW, for exterior girders is

TABLE 3.1

Dead Load Summary of Unfactored Shears and Moments for Interior and Exterior Girders

Girder Location	V_{DC} (kips)	M_{DC} (ft-kips)	V_{DW} (kips)	M_{DW} (ft-kips)
Interior	46.0	690.3	9.0	135.0
Exterior	40.0	600.3	9.0	135.0

$$M_{DW} = \frac{wL^2}{8} = \frac{\left(0.3\,\frac{\text{kips}}{\text{ft}}\right)(60\text{ ft})^2}{8} = 135 \text{ ft-kips}$$

Please see Table 3.1.

Step 3: Evaluate Live Loads

The number of lanes (with fractional parts discounted) is

$$N_L = \frac{w}{12} = \frac{43\text{ ft}}{12\,\frac{\text{ft}}{\text{lane}}} = 3.58 \text{ lanes (3 lanes)}$$

Find the longitudinal stiffness parameter, K_g.

A Art. 4.6.2.2.1

The modulus of elasticity for the concrete deck is

A Eq. 5.4.2.4-1

$$E_{deck} = E_c = 33{,}000\, w_c^{1.5}\sqrt{f_c'} = (33{,}000)\left(0.15\,\frac{\text{kips}}{\text{ft}^3}\right)\sqrt{5\text{ ksi}} = 4286.8 \text{ kips/in}^2$$

The modulus of elasticity for the beam, E_{beam}, is 29,000 kips/in^2.
The modular ratio between steel and concrete is

$$n = \frac{E_{beam}}{E_{deck}} \frac{29{,}000\text{ ksi}}{4286.8\text{ ksi}} = 6.76$$

The distance between the centers of gravity of the deck and beam is (see Figure 3.5)

$$e_g = 19.7\text{ in} + 2\text{ in} + 4\text{ in} = 25.7\text{ in}$$

The moment of inertia for the beam, I (steel section only), is 19,600 in^4.

The area, A, is 73.3 in^2.

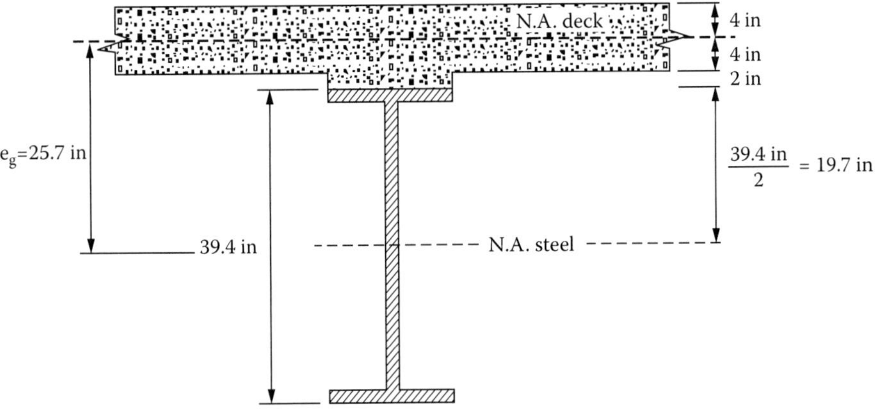

FIGURE 3.5
Section for longitudinal stiffness parameter, K_g.

The longitudinal stiffness parameter is

$$K_g = n(I + Ae_g^2) = (6.76)(19{,}600\ \text{in}^4 + (73.3\ \text{in}^3)(25.7\ \text{in})^2)$$

$$= 459.77\ \text{in}^4$$

A Art. 4.6.2.2.1; A Eq. 4.6.2.2.1-1

The multiple presence factors have been included in the approximate equations for the live distribution factors. These factors must be applied where the lever rule is specified.

A Tbl. 3.6.1.1.2-1; Comm. C3.6.1.1.2

The distribution factor for moments for interior girders of cross-section type (a) with one design lane loaded is

A Tbl. 4.6.2.2.2b-1 or Appendix A

$$\text{DFM}_{\text{si}} = 0.06 + \left(\frac{S}{14}\right)^{0.4}\left(\frac{S}{L}\right)^{0.3}\left(\frac{K_g}{12\ Lt_s^3}\right)^{0.1}$$

$$= 0.06 + \left(\frac{10\ \text{ft}}{14}\right)^{0.4}\left(\frac{10\ \text{ft}}{60\ \text{ft}}\right)^{0.3}\left(\frac{459{,}774\ \text{in}^4}{12(60\ \text{ft})(7.5\ \text{in})^3}\right)^{0.1}$$

$$= 0.59$$

where:
S = spacing of beams (ft)
L = span length (ft)
t_s = depth of concrete slab (in)

The distribution factor for moments for interior girders with two or more design lanes loaded is

$$DFM_{mi} = 0.075 + \left(\frac{S}{9.5}\right)^{0.6}\left(\frac{S}{L}\right)^{0.2}\left(\frac{K_g}{12\,Lt_s^3}\right)^{0.1}$$

$$= 0.075 + \left(\frac{10\text{ ft}}{9.5}\right)^{0.6}\left(\frac{10\text{ ft}}{60\text{ ft}}\right)^{0.2}\left(\frac{459{,}774\text{ in}^4}{12(60\text{ ft})(7.5\text{ in})^3}\right)^{0.1}$$

$$= 0.82 \text{ [controls]}$$

The distribution factor for shear for interior girders of cross-section type (a) with one design lane loaded is

A Tbl. 4.6.2.2.3a-1 or Appendix C

$$DFV_{si} = 0.36 + \frac{S}{25} = 0.36 + \frac{10\text{ ft}}{25} = 0.76$$

The distribution factor for shear for interior girders with two or more design lanes loaded is

$$DFV_{mi} = 0.2 + \frac{S}{12} - \left(\frac{S}{35}\right)^2 = 0.2 + \frac{10\text{ ft}}{12} - \left(\frac{10\text{ ft}}{35}\right)^2 = 0.952 \text{ [controls]}$$

The distribution factors for exterior girders of typical cross section (a).

A Tbl. 4.6.2.2.1-1, 4.6.2.2.2d-1

Use the lever rule to find the distribution factor for moments for exterior girders with one design lane loaded. See Figure 3.6.

A Art. 3.6.1.3.1

$$\Sigma M_{@hinge} = 0$$

$$0 = \frac{P}{2}(9.5\text{ ft}) + \frac{P}{2}(3.5\text{ ft}) - R(10\text{ ft})$$

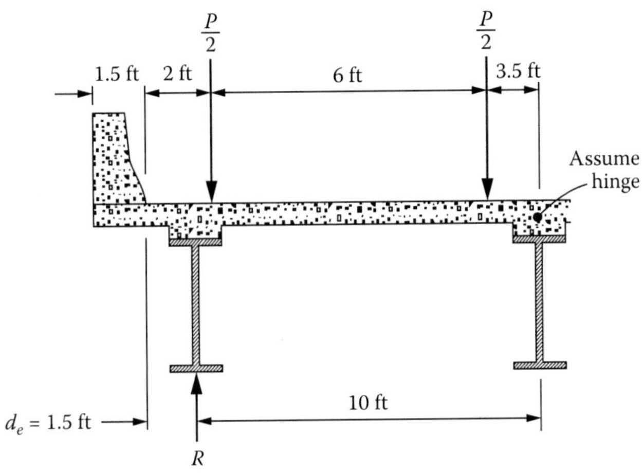

FIGURE 3.6
Lever rule for the distribution factor for moments for exterior girder.

$$R = 0.65\ P$$

$$R = 0.65$$

The distribution factor for moments for exterior girders with one design lane loaded is

A Tbl. 3.6.1.1.2-1

$$DFM_{se} = (R)(\text{multiple presence factor for one lane loaded, m})$$

$$= (0.65)(1.2) = 0.78\ [\text{controls}]$$

The distribution factor for moments for exterior girders with two or more design lanes loaded, g, is

A Tbl. 4.6.2.2.2d-1 or Appendix B

$$g = eg_{interior}$$

where:
$g_{interior}$ = distribution factor for interior girder = DFM_{mi}
g = DFM_{me} = $e(DFM_{mi})$

The distance from the center of the exterior web of exterior beam to the interior edge of the barrier, d_e, is 1.5 ft.

A Tbl. 4.6.2.2.2d-1

The correction factor for distribution is

$$e = 0.77 + \frac{d_e}{9.1} = 0.77 + \frac{1.5\ \text{ft}}{9.1} = 0.935$$

The distribution factor for moments for exterior girders with two or more design lanes loaded is

$$DFM_{me} = e(DFM_{mi}) = (0.935)(0.826) = 0.77$$

Use the lever rule to find the distribution factor for shear for exterior girder with one design lane loaded.

A Tbl. 4.6.2.2.3b-1 or Appendix D

This is the same as DFM_{se} for exterior girders with one design lane loaded.

$$DFV_{se} = DFM_{se} = 0.78 \text{ [controls]}$$

Find the distribution factor for shear for exterior girders with two or more design lanes loaded using g:

$$g = (e)(g_{int})$$

$$g = DFV_{me} = (e)(DFV_{mi})$$

where:
g_{int} = distribution factor for interior girder = DFV_{mi}

The correction factor for distribution is

A Tbl. 4.6.2.2.3b-1 or Appendix D

$$e = 0.6 + \frac{d_e}{10} = 0.6 + \frac{1.5\ \text{ft}}{10} = 0.75$$

The distribution factor for shear for exterior girders with two or more design lanes loaded is

$$DFV_{me} = (0.75)(0.952) = 0.714$$

The HL-93 live load is made of the design lane load plus either the design truck (HS-20) or the design tandem load (whichever is larger).

A Art. 3.6.1.2

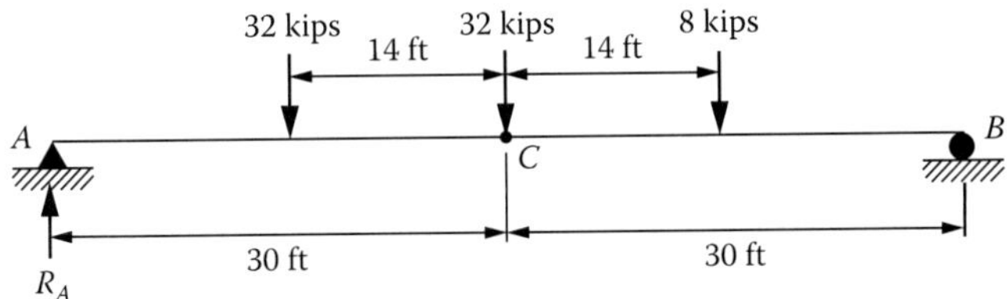

FIGURE 3.7
Design truck (HS-20) load moment at midspan.

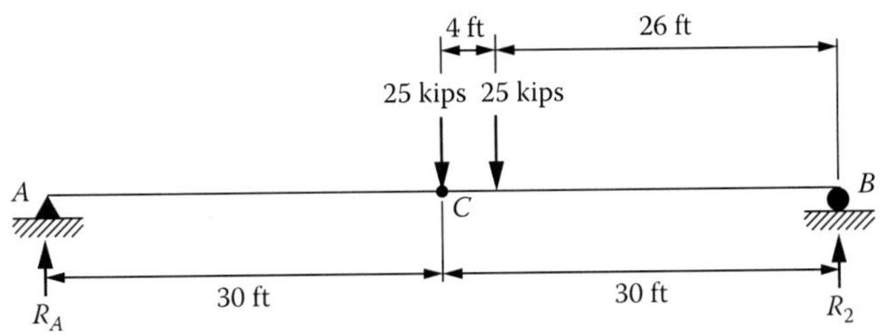

FIGURE 3.8
Design tandem load moment at midspan.

Find the design truck (HS-20) moment due to live load. Please see Figure 3.7.

$$\Sigma M_{@B} = 0$$

The reaction at A is

$$R_A(60\text{ ft}) = (32\text{ kips})(30\text{ ft} + 14\text{ ft}) + (32\text{ kips})(30\text{ ft}) + (8\text{ kips})(30\text{ ft} - 14\text{ ft})$$

$$R_A = 41.6\text{ kips}$$

The design truck moment due to live load is

$$M_{tr} = (41.6\text{ kips})(30\text{ ft}) - (32\text{ kips})(14\text{ ft}) = 800\text{ ft-kips [controls]}$$

Find the design tandem moment. See Figure 3.8.

$$\Sigma M_{@B} = 0$$

The reaction at A is

$$R_A(60\text{ ft}) = (25\text{ kips})(30\text{ ft}) + (25\text{ kips})(26\text{ ft})$$

$$R_A = 23.33\text{ kips}$$

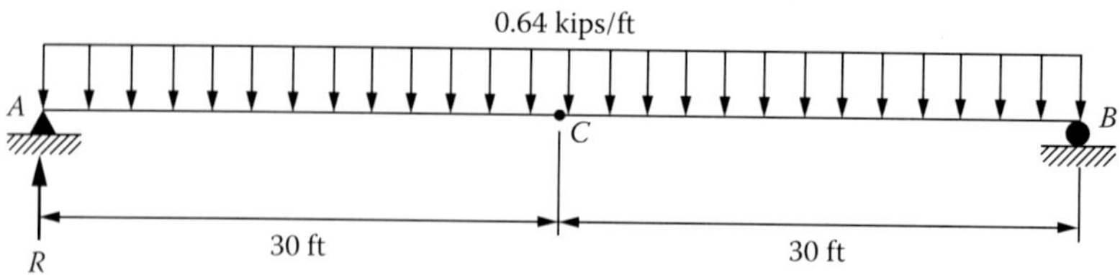

FIGURE 3.9
Design lane load moment.

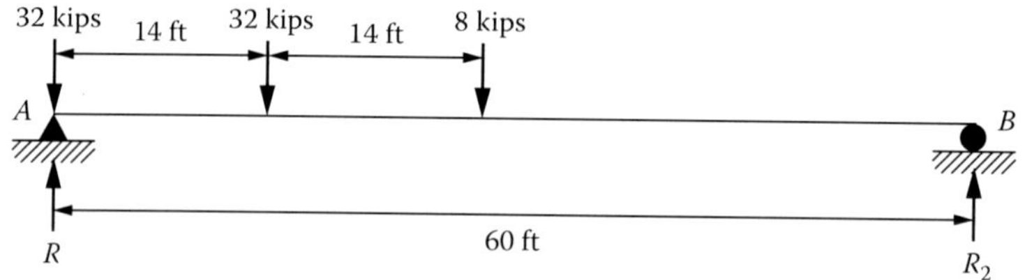

FIGURE 3.10
Design truck (HS-20) shear at support.

The design tandem moment is

$$M_{tandem} = (23.33 \text{ kips})(30 \text{ ft}) = 699.9 \text{ ft-kips } (700 \text{ ft-kips})$$

Find the design lane moment. Please see Figure 3.9.

$$M_c = M_{lane} = \frac{wL^2}{8} = \frac{\left(0.64 \frac{\text{kips}}{\text{ft}}\right)(60 \text{ ft})^2}{8} = 288 \text{ ft-kips}$$

The dynamic load allowance, IM, is 33% (applied to design truck or design tandem only, not to the design lane load).

A Art. 3.6.2.1

The unfactored total live load moment per lane is

$$M_{LL+IM} = M_{tr}(1 + IM) + M_{ln}$$

$$= (800 \text{ ft-kips})(1 + 0.33) + 288 \text{ ft-kips} = 1352 \text{ ft-kips per lane}$$

Find the design truck shear. See Figure 3.10.

$$\Sigma M_{@B} = 0$$

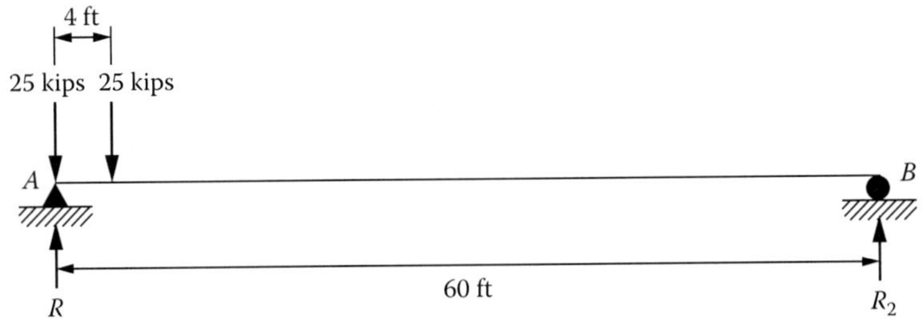

FIGURE 3.11
Design tandem load shear at support.

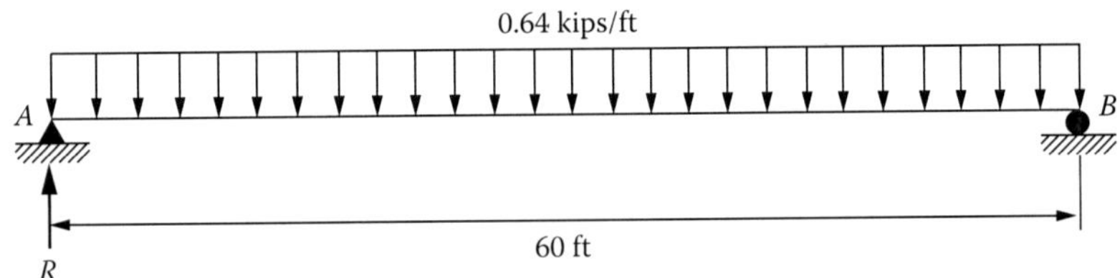

FIGURE 3.12
Design lane load shear.

The reaction at A is

$$R_A(60\text{ ft}) = (32\text{ kips})(60\text{ ft}) + (32\text{ kips})(46\text{ ft}) + (8\text{ kips})(32\text{ ft})$$

$$R_A = 60.8\text{ kips}$$

The design truck shear, V_{tr}, is 60.8 kips [controls].
Find the design tandem shear. See Figure 3.11.

$$\Sigma M_{@B} = 0$$

The reaction at A is

$$R_A(60\text{ ft}) = (25\text{ kips})(60\text{ ft}) + (25\text{ kips})(56\text{ ft})$$

$$R_A = 48.33\text{ kips}$$

The design tandem shear, V_{tandem}, is 48.33 kips.
Find the design lane shear. See Figure 3.12.

$$V_{ln} = \frac{wL}{2} = \frac{\left(0.64\dfrac{\text{kips}}{\text{ft}}\right)(60\text{ ft})}{2} = 19.2\text{ kips}$$

TABLE 3.2 (A Art. 4.6.2.2.2)

Summary of Live Load Effects*

Girder Location	No. of Lanes Loaded	Unfactored M_{LL+IM} (ft-kips per Lane)	DFM	Unfactored M_{LL+IM} (ft-kips per Girder)	Unfactored V_{LL+IM} (kips per Lane)	DFV	Unfactored V_{LL+IM} (kips per Girder)
Interior	1	1352.0	0.596		100.1	0.76	—
	2 or more	1352.0	0.82	1108.6	100.1	0.952	95.3
Exterior	1	1352.0	0.78	1054.6	100.1	0.78	78.1
	2 or more	1352.0	0.77		100.1	0.714	—

* The live load moments and shears per girder are determined by applying the distribution factors, DF, to those per lane.

The dynamic load allowance, IM, is 33% (applied to design truck or design tandem only, not to the design lane load).

A Art. 3.6.2.1

The unfactored total live load shear is

$$V_{LL+IM} = V_{tr}(1 + IM) + V_{ln}$$

$$= (60.8\text{ kips})(1 + 0.33) + 19.2\text{ kips} = 100.1\text{ kips}$$

Please see Table 3.2.

Total live load moment and shear for interior girders

$$M_{LL+IM} = 1108.6\text{ ft-kips}$$

$$V_{LL+IM} = 95.3\text{ kips per girder}$$

Total live load moment and shear for exterior girders

$$M_{LL+IM} = 1054.6\text{ ft-kips}$$

$$V_{LL+IM} = 78.1\text{ kips}$$

Step 4: Check the Strength I Limit State.

A Arts. 3.3.2; 3.4; Tbls. 3.4.1-1, 3.4.1-2

$$U = 1.25(DC) + 1.5(DW) + 1.75(LL + IM)$$

The factored moment for interior girders is

$$M_u = 1.25\,M_{DC} + 1.5\,M_{DW} + 1.75\,M_{LL+IM}$$

$$= (1.25)(690.3\text{ ft-kips}) + (1.5)(135.0\text{ ft-kips}) + (1.75)(1108.6\text{ ft-kips})$$

$$= 3005.4\text{ ft-kips}$$

The factored shear for interior girders is

$$V_u = 1.25\ V_{DC} + 1.5\ V_{DW} + 1.75\ V_{LL+IM}$$
$$= (1.25)(46\ \text{kips}) + (1.5)(9.0\ \text{kips}) + (1.75)(95.3\ \text{kips})$$
$$= 237.8\ \text{kips}\ (238.0\ \text{kips})$$

The factored moment for exterior girders is

$$M_u = (1.25)(600.3\ \text{ft-kips}) + (1.5)(135\ \text{ft-kips}) + (1.75)(1054.6\ \text{ft-kips})$$
$$= 2798.4\ \text{ft-kips}\ (2800\ \text{ft-kips})$$

The factored shear for exterior girders is

$$V_u = (1.25)(40\ \text{kips}) + (1.5)(9.0\ \text{kips}) + (1.75)(78.1\ \text{kips})$$
$$= 200.2\ \text{kips}\ (200\ \text{kips})$$

Check the Flexural Resistance.

Find the flange stress, f_{bu}, for noncomposite sections with continuously braced flanges in tension or compression. The required flange stress without the flange lateral bending, f_{bu}, must satisfy the following.

A Art. 6.10.8.1.3

$$f_{bu} \le \Phi_f R_h F_{yf}$$

A Eq. 6.10.8.1.3-1

R_h hybrid factor — 1.0 (for rolled shapes)

A Art. 6.10.1.10.1

Φ_f resistance factor for flexure — 1.0

A Art. 6.5.4.2

F_{yf} minimum yield strength of a flange — 50 ksi

The flange stress for interior girders is

$$f_{bu} = \frac{M_u}{S_x} = \frac{(3005.4\ \text{ft-kips})\left(12\frac{\text{in}}{\text{ft}}\right)}{992\ \text{in}^3}$$

$$= 36.4\ \text{kips/in}^2 < \Phi_f R_h F_{yf} = 50\ \text{kips/in}^2\ \text{[OK]}$$

Check the shear resistance in the web due to the factored loads, V_u, where:

A Arts. 6.10.9.1, 6.10.9.2

V_p = plastic shear force
C = ratio of the shear–buckling resistance to the shear yield strength

A Art. 6.10.9.3.2

$$V_u \le \Phi V_n \text{ and } V_n = CV_p$$

Shear in the web, V_u, must satisfy the following.

$$V_u \le \Phi_v V_n$$

A Eq. 6.10.9.1-1

Nominal shear resistance,

A Eq. 6.10.9.2-1

$V_n = CV_p$

F_{yw}	specified minimum yield strength of a web	50 ksi
t_w	web thickness	0.75 in
E	modulus of elasticity of steel	29,000 ksi
Φ_v	resistance factor for shear	1.00

A Art. 6.5.4.2

D	depth of steel beam	39.4 in

The shear yielding of the web is

A Eq. 6.10.9.2-2

$$V_p = 0.58\ F_{yw}Dt_w$$

$$= (0.58)(50\text{ ksi})(39.4\text{ in})(0.75\text{ in})$$

$$= 856.95\text{ kips (857 kips)}$$

Find the ratio of the shear buckling resistance to the shear yield strength, C.
C is determined from any of these AASHTO equations: Eq. 6.10.9.3.2-4, 6.10.9.3.2-5, 6.10.9.3.2-6.

A Art. 6.10.9.3.2

Try AASHTO Eq. 6.10.9.3.2-4.

$$\text{If } \frac{D}{t_w} \le 1.12\sqrt{\frac{Ek}{F_{yw}}}\text{, then C = 1.0.}$$

The shear buckling coefficient, k, is 5 for unstiffened web panels.

A Com. 6.10.9.2

$$\frac{D}{t_w} = \frac{39.4 \text{ in}}{0.75 \text{ in}} = 52.5$$

$$1.12\sqrt{\frac{Ek}{F_{yw}}} = 1.12\sqrt{\frac{(29{,}000 \text{ ksi})(5)}{50 \text{ ksi}}} = 60.3 > \frac{D}{t_w} = 52.5 \text{ [OK]}$$

Therefore, C is 1.0.

The nominal shear resistance for unstiffened webs is

A Eq. 6.10.9.2-1

$$V_n = CV_p = (1.0)(857 \text{ kips}) = 857 \text{ kips [OK]}$$

The shear resistance for interior girders is greater than the factored shear for interior girder.

$$\Phi_v V_n = (1.0)(857 \text{ kips}) = 857 \text{ kips} > V_u = 238 \text{ kips [OK]}$$

A Art. 6.10.9

Step 5: Check the Fatigue Limit State

Details shall be investigated for fatigue as specified in AASHTO Art. 6.6.1. The fatigue load combination in AASHTO Tbl. 3.4.1-1 and the fatigue live load specified in AASHTO Art. 3.6.1.4 shall apply.

A Art. 6.10.5.1

For the load induced fatigue, each detail must satisfy the following:

A Art. 6.6.1.2, A Eq. 6.6.1.2.2-1

$$\gamma(\Delta f) \leq (\Delta F)_n$$

γ	load factor	
(Δf)	live load stress range due to fatigue load	ksi
$(\Delta F)_n$	nominal fatigue resistance as specified in Art. 6.6.1.2.5	ksi

The fatigue load is one design truck with a constant spacing of 30 ft between 32 kip axles. Find the moment due to fatigue load.

A Art. 3.6.1.4

Find the center of gravity of design truck, x, from the left 32 kip axle

$$x = \frac{(32 \text{ kips})(30 \text{ ft}) + (8 \text{ kips})(30 \text{ ft} + 14 \text{ ft})}{32 \text{ kips} + 32 \text{ kips} + 8 \text{ kips}}$$

$$= 18.22 \text{ ft from left 32 kip axle}$$

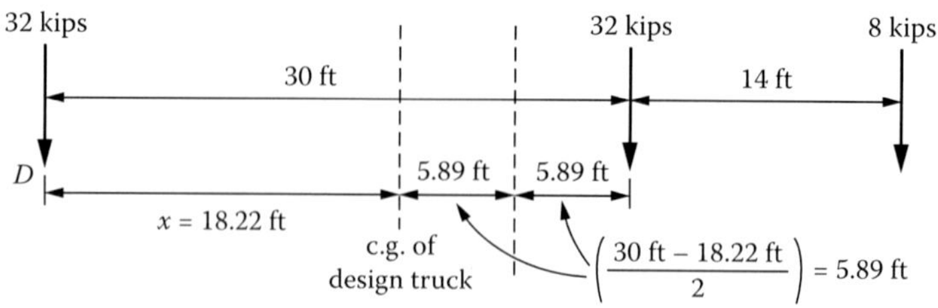

FIGURE 3.13
Center of gravity of design truck loading (HS-20).

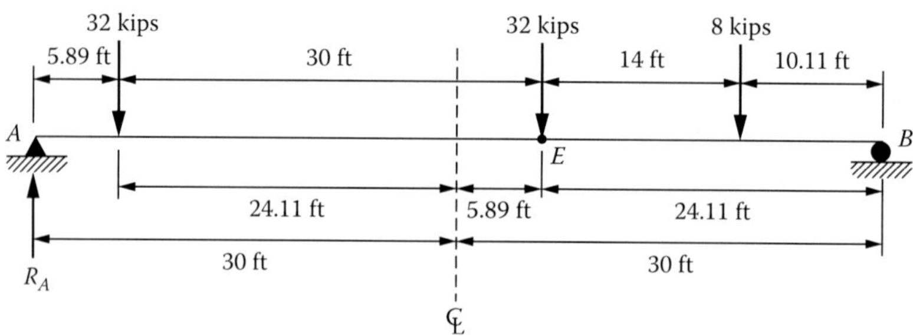

FIGURE 3.14
Fatigue load position for maximum moment.

Please see Figures 3.13 and 3.14.

$$\Sigma M_{@B} = 0$$

The reaction at A is

$$R_A(60 \text{ ft}) = (32 \text{ kips})(54.11 \text{ ft}) + (32 \text{ kips})(24.11 \text{ ft}) + (8 \text{ kips})(10.11 \text{ ft})$$

$$R_A = 43.07 \text{ kips}$$

The maximum moment at point E,

$$M_E = (43.07 \text{ kips})(35.89 \text{ ft}) - (32 \text{ kips})(30 \text{ ft}) = 585.8 \text{ ft-kips [controls]}$$

The maximum moment due to fatigue load is 585.8 ft-kips.

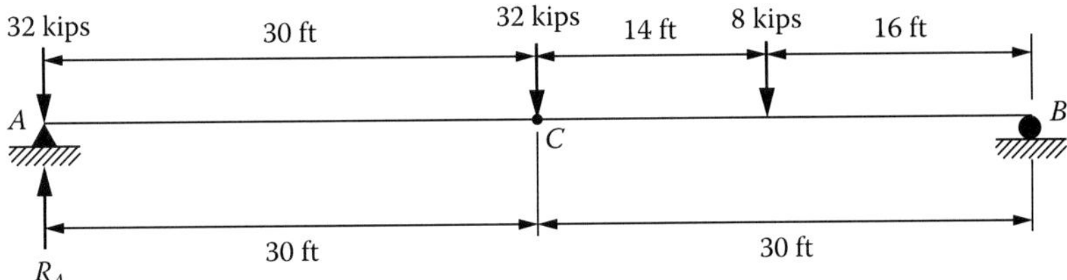

FIGURE 3.15
Fatigue load position for maximum shear.

Find the shear due to the Fatigue Load, V_{fat}. See Figure 3.15.

$$\Sigma M_{@B} = 0$$

The reaction at A is

$$R_A = (32\text{ kips}) + \left(\frac{32\text{ kips}}{2}\right) + (8\text{ kips})\left(\frac{16\text{ ft}}{60\text{ ft}}\right)$$

$$= 50.13\text{ kips}$$

$$V_{fat} = 50.13\text{ kips}$$

The dynamic load allowance, IM, is 15%.

A Tbl. 3.6.2.1-1

Find the Load Distribution for Fatigue, DFM_{fat}.

The distribution factor for one traffic lane shall be used. For single lane, approximate distribution factors in AASHTO Art. 4.6.2.2 are used. The force effects shall be divided by 1.20.

A Art. 3.6.1.4.3b, 3.6.1.1.2

The distribution factor for fatigue moments in interior girders with one lane is

$$DFM_{fat,int} = \frac{DFM_{int}}{1.2} = \frac{0.59}{1.2} = 0.49$$

The distribution factor for the fatigue moments in exterior girders with one loaded lane is

$$DFM_{fat,ext} = \frac{DFM_{ext}}{1.2} = \frac{0.78}{1.2} = 0.65$$

The unfactored distributed moment for interior girders is

$$M_{fat,int} = M_{fatigue}(DFM_{fat,int})(1 + IM)$$

$$= (585.8 \text{ ft-kips})(0.49)(1 + 0.15) = 330.1 \text{ ft-kips } (330 \text{ ft-kips})$$

The unfactored distributed moment for exterior girders is

$$M_{fat,ext} = M_{fatigue}(DFM_{fat,ext})(1 + IM)$$

$$= (585.8 \text{ ft-kips})(0.65)(1 + 0.15) = 437.9 \text{ ft-kips } (438 \text{ ft-kips})$$

Live load stress due to fatigue load for interior girders is

$$(\Delta f_{int}) = \frac{M_{fat,int}}{S_x} = \frac{330.0 \text{ ft-kips}\left(12 \frac{\text{in}}{\text{ft}}\right)}{992 \text{ in}^3} = 3.99 \text{ kips/in}^2$$

Live load stress due to fatigue load for exterior girders is

$$(\Delta f_{ext}) = \frac{M_{fat,ext}}{S_x} = \frac{438 \text{ ft-kips}\left(12 \frac{\text{in}}{\text{ft}}\right)}{992 \text{ in}^3} = 5.30 \text{ kips/in}^2$$

For load-induced fatigue, each detail shall satisfy:

A Art. 6.6.1.2.2

$$\gamma(\Delta f) \le (\Delta F)_n$$

Find the nominal fatigue resistance, $(\Delta F)_n$, for Fatigue II load combination for finite life.

A Art. 6.6.1.2.5, A Eq. 6.6.1.2.5-2

$$(\Delta F)_n = \left(\frac{A}{N}\right)^{1/3}$$

The girder is in Detail Category A, because it is a plain rolled member.

A Tbl. 6.6.1.2.3-1

A = constant taken for Detail Category A $\quad 250 \times 10^8$ kips/in²

A Tbl. 6.6.1.2.5-1

n = number of stress range cycles per truck passage	1.0 (for simple span girders with span greater than 40 ft) **A Tbl. 6.6.1.2.5-2**
p = the fraction of truck traffic in a single lane	0.80 (because N_L is 3 lanes) **A Tbl. 3.6.1.4.2-1**
$(\Delta F)_{TH}$ = the constant amplitude threshold for Detail Category A	24.0 kips/in^2 **A Tbl. 6.6.1.2.5-3**

$ADTT_{SL}$ = the number of trucks per day in a single lane over the design life
ADTT = the number of trucks per day in one direction over the design life

$$ADTT_{SL} = (p)(ADTT) = (0.80)(250) = 200 \text{ trucks daily}$$

A Art. 3.6.1.4.2

The number of cycles of stress range N is

$$N = (365 \text{ days})(75 \text{ years})n(ADTT)_{SL} = (365 \text{ days})(75 \text{ years})(1.0)(200 \text{ trucks})$$

$$= 5{,}475{,}000 \text{ cycles}$$

A Eq. 6.6.1.2.5-3

The nominal fatigue resistance is

A Eq. 6.6.1.2.5-2

$$(\Delta F)_n = \left(\frac{A}{N}\right)^{1/3} = \left(\frac{250 \times 10^8 \dfrac{\text{kips}}{\text{in}^2}}{5{,}475{,}000 \text{ cycles}}\right)^{1/3} = 16.6 \text{ kips/in}^2$$

Find the Load Factor, γ, and confirm that the following are satisfied.

A Eq. 6.6.1.2.2-1

$$\gamma(\Delta f) \le (\Delta F)_n$$

γ = 0.75 for Fatigue II Limit State

A Tbl. 3.4.1-1

The factored live load stress due to the fatigue load for interior girders is

$$\gamma(\Delta f_{int}) = (0.75)(3.99 \text{ ksi}) = 2.99 \text{ ksi} < 16.6 \text{ ksi [OK]}$$

The factored live load stress due to the fatigue load for exterior girders is

$$\gamma(\Delta f_{ext}) = (0.75)(5.30 \text{ ksi}) = 3.98 \text{ ksi} < 16.6 \text{ ksi [OK]}$$

A Art. 6.10.5.3

Special Fatigue Requirement for Webs

Find the shear in the web due to unfactored permanent load plus factored fatigue shear load, V_u, for interior and exterior girders, satisfying the following provision to control web-buckling and elastic flexing of the web. V_{cr} is the shear buckling resistance (Eq. 6.10.9.3.3-1). The use of the Fatigue I load provisions for the Fatigue II load combination will be considered conservative.

A Eq. 6.10.9.3.3-1

The distribution factor for fatigue shear in interior girders with one design lane loaded is

$$DFV_{fat,int} = \frac{DFV_{int}}{m} = \frac{0.76}{1.2} = 0.63$$

The distribution factor for fatigue shear in exterior girders with one design lane loaded is

$$DFV_{fat,ext} = \frac{DFV_{ext}}{m} = \frac{0.78}{1.2} = 0.65$$

Unfactored fatigue shear load per interior girder is

$$(DFV_{fat,int})(V_{fat}) = (0.63)(50.13 \text{ kips}) = 31.58 \text{ kips}$$

Unfactored fatigue shear load per exterior girder is

$$(DFV_{fat,ext})(V_{fat}) = (0.65)(50.13 \text{ kips}) = 32.58 \text{ kips}$$

$$V_u \le V_{cr}$$

A Eq. 6.10.5.3-1

$$V_u = V_{DL} + \gamma(V_{fat})(1 + IM)$$

The shear buckling resistance for unstiffened webs, V_{cr}, for W40 × 249 with F_y equal to 50 ksi is

A Art. 6.10.9.2; Eq. 6.10.9.2-1

$$V_{cr} = CV_p$$

$$C = 1.0$$

A Eq. 6.10.9.3.2-4

$$V_p = 0.58\, F_{yw}Dt_w = 857 \text{ kips}$$

A Eq. 6.10.9.2-2

$$V_{cr} = CV_p = (1.0)(857 \text{ kips}) = 857 \text{ kips}$$

The distortion-induced fatigue for interior girders is

$$V_{DL} = V_{DC} + V_{DW}$$

$$V_u = V_{DL} + \gamma V_{fat}(1 + IM) = (46 \text{ kips} + 9 \text{ kips}) + (0.75)(31.58 \text{ kips})(1.15)$$

$$= 82.23 \text{ kips} < V_{cr} = 857 \text{ kips[OK]}$$

The distortion induced fatigue for exterior girders is

$$V_u = V_{DL} + \gamma V_{fat}(1 + IM) = (40 \text{ kips} + 9 \text{ kips}) + (0.75)(32.58 \text{ kips})(1.15)$$

$$= 77.1 \text{ kips} < V_{cr} = 857 \text{ kips[OK]}$$

Step 6: Check the Service Limit States

Deflection Limit

A Art. 2.5.2.6

The maximum deflection for vehicular load, δ_{max}, where L is the span, is

A Art. 2.5.2.6.2

$$\delta_{max} = \frac{L}{800}$$

Deflection will be the larger of

A Art. 3.6.1.3.2

1. Deflection resulting from the design truck (HS-20) alone
2. Deflection resulting from 25% of design truck plus the design lane load

Find the deflection resulting from the design truck alone (see Figure 3.16)

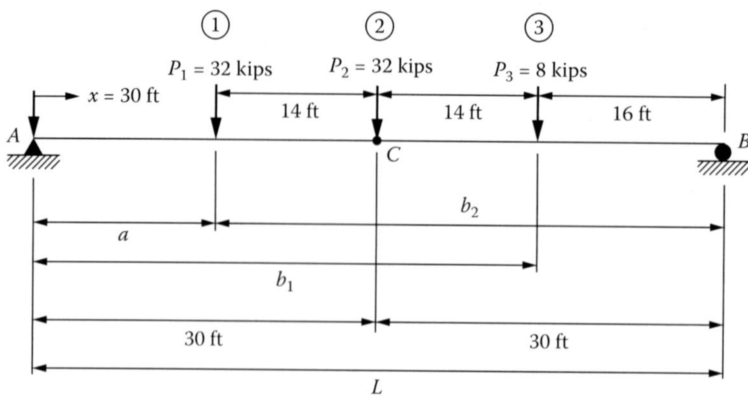

FIGURE 3.16
Design truck loading for maximum deflection at midspan.

The deflections at midspan, δ_1, δ_2, δ_3, are

$$\delta_1 = \frac{P_1 b_2 x}{6\ EIL}\left(L^2 - b_2^2 - x^2\right)$$

$$= \frac{(32\ \text{kips})(30\ \text{ft} + 14\ \text{ft})(30\ \text{ft})\left(12\frac{\text{in}}{\text{ft}}\right)^2}{(6)(29{,}000\ \text{ksi})(19{,}600\ \text{in}^4)(60\ \text{ft})\left(12\frac{\text{in}}{\text{ft}}\right)}$$

$$\left(\left(60\ \text{ft}\left(12\frac{\text{in}}{\text{ft}}\right)\right)^2 - \left((30\ \text{ft} + 14\ \text{ft})\left(12\frac{\text{in}}{\text{ft}}\right)\right)^2 - \left(30\ \text{ft}\left(12\frac{\text{in}}{\text{ft}}\right)\right)^2\right)$$

$$= 0.273\ \text{in}$$

$$\delta_2 = \frac{P_2 L^3}{48\ EI} = \frac{(32\ \text{kips})(60\ \text{ft})^3\left(12\frac{\text{in}}{\text{ft}}\right)^3}{(48)(29{,}000\ \text{ksi})(19{,}600\ \text{in}^4)}$$

$$= 0.438\ \text{in}$$

$$\delta_3 = \frac{P_3 b_1 x}{6\,EIL}\left(L^2 - b_1^2 - x^2\right)$$

$$= \frac{(8\text{ kips})(30\text{ ft} + 14\text{ ft})(30\text{ ft})\left(12\dfrac{\text{in}}{\text{ft}}\right)^2}{(6)(29{,}000\text{ ksi})(19{,}600\text{ in}^4)(60\text{ ft})\left(12\dfrac{\text{in}}{\text{ft}}\right)}$$

$$\left(\left(60\text{ ft}\left(12\frac{\text{in}}{\text{ft}}\right)\right)^2 - \left((30\text{ ft} + 14\text{ ft})\left(12\frac{\text{in}}{\text{ft}}\right)\right)^2 - \left(30\text{ ft}\left(12\frac{\text{in}}{\text{ft}}\right)\right)^2\right)$$

$$= 0.068\text{ in}$$

$$\delta_{LL} = \delta_1 + \delta_2 + \delta_3 = 0.273\text{ in} + 0.438\text{ in} + 0.068\text{ in} = 0.779\text{ in per lane}$$

m multiple presence factor for three loaded lanes 0.85

A Tbl. 3.6.1.1.2-1

N_L number of lanes 3
N_g number of girders 5

The dynamic load allowance, IM, is 33%.

A Art. 3.6.2.1

Distribution factor for deflection is equal to the number of lanes, N_L, divided by the number of beams, N_g.

A Comm. 2.5.2.6.2

$$DF = (m)\left(\frac{N_L}{N_g}\right)$$

A Tbl. 3.6.2.1-1

$$DF = (0.85)\left(\frac{3}{5}\right) = 0.51$$

$$IM = 33\%$$

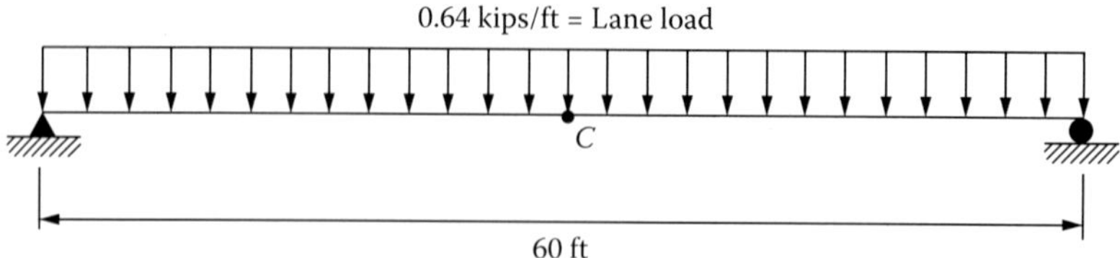

FIGURE 3.17
Design lane loading for maximum deflection at midspan.

The distributed midspan deflection per girder due to the design truck alone is

$$\delta_{LL+IM} = (0.779 \text{ in per lane})(0.51)(1 + 0.33) = 0.53 \text{ in per girder [controls]}$$

A Art. 3.6.1.3.2

Find the deflection resulting from 25% of design truck plus the lane load.

A Art. 3.6.1.3.2

The deflection at midspan due to the lane load is (see Figure 3.17)

$$\delta_{ln} = \frac{5\,wL^4}{384\,EI} = \frac{(5)\left(0.64\,\frac{\text{kips}}{\text{ft}}\right)\left(\frac{\text{ft}}{12\text{ in}}\right)(60\text{ft})\left(\frac{12\text{ in}}{\text{ft}}\right)^4}{(384)(29{,}000\text{ ksi})(19{,}600\text{ in}^4)} = 0.33\text{ in per lane}$$

The midspan deflection per lane is

$$\delta_{LL25\%+ln} = 25\% \text{ of deflection due to design truck} + \delta_{ln}$$

$$= (0.25)(\delta_{LL}) + \delta_{ln}$$

$$= (0.25)(0.779\text{ in}) + 0.33\text{ in} = 0.525\text{ in per lane [controls]}$$

The distributed midspan deflection per girder is

$$\delta_{LL+IM} = (\delta_{LL25\%+ln})(DF)(1 + IM) = (0.525\text{ in per lane})(0.51)(1 + 0.33)$$

$$= 0.356\text{ in per girder}$$

Confirm that $\delta_{LL+IM} < \delta_{max}$.

$$\delta_{max} = \frac{L}{800} = \frac{(60\ \text{ft})\left(12\frac{\text{in}}{\text{ft}}\right)}{800} = 0.90\ \text{in}$$

$$\delta_{LL+IM} = 0.525\ \text{in} < \delta_{max} = 0.90\ \text{in [OK]}$$

A Art. 6.10.4.2

Permanent deformations

For inelastic deformation limitations the Service II Limit State load combination shall apply and both steel flanges of noncomposite sections shall satisfy the following.

Flexure (to prevent objectionable permanent deflections)

A Art. 6.10.4.2.2

Both steel flanges of noncomposite sections shall satisfy the following, where f_f is the flange stress due to Service II loads.

A Eq. 6.10.4.2.2-3

$$f_f + \frac{f_l}{2} \le 0.80\ R_h F_{yf}$$

f_l	flange lateral bending stress due to Service II loads	0 (for continually braced flanges)
R_h	hybrid factor	1.0 (for rolled shapes)

A Art. 6.10.1.10.1

Service II Limit State: U = 1.0 DC + 1.0 DW + 1.3(LL + IM)

A Tbl. 3.4.1-1

The factored design moment for interior girders is

$$M_u = 1.0\ M_{DC} + 1.0\ M_{DW} + 1.3\ M_{LL+IM}$$

$$= (1.0)(690.3\ \text{ft-kips}) + (1.0)(135\ \text{ft-kips} + (1.3)(1108.6\ \text{ft-kips}))$$

$$= 2266.5\ \text{ft-kips}$$

The flange stress due to Service II loads for interior girders is

$$f_f = \frac{M_u}{S_x} = \frac{2266.5\ \text{ft-kips}\left(12\frac{\text{in}}{\text{ft}}\right)}{992\ \text{in}^3} = 27.4\ \text{kips/in}^2$$

For steel flanges of noncomposite sections in interior girders, check,

A Art. 6.10.4.2.2, A Eq. 6.10.4.2.2-3

$$f_f + \frac{f_l}{2} \le 0.80\, R_h F_{yf}$$

$$27.4 \text{ ksi} + \frac{0}{2} = 27.4 \text{ ksi} < (0.8)(1.0)(50 \text{ ksi}) = 40 \text{ ksi} \quad \text{[OK]}$$

The factored design moment for exterior girders is

$$M_u = (1.0)(600.3 \text{ ft-kips}) + (1.0)(135 \text{ ft-kips}) + (1.3)(1054.6 \text{ ft-kips})$$

$$= 2106.8 \text{ ft-kips}$$

The flange stress due to Service II loads for exterior girders is

$$f_f = \frac{M_u}{S_x} = \frac{(2106.8 \text{ ft-kips})\left(12 \frac{\text{in}}{\text{ft}}\right)}{992 \text{ in}^3} = 25.5 \text{ ksi}$$

For steel flanges of noncomposite section in exterior girders, check,

$$f_f + \frac{f_l}{2} \le 0.80\, R_h F_{yf}$$

$$25.5 \text{ ksi} + \frac{0}{2} = 25.5 \text{ ksi} < (0.8)(1.0)(50 \text{ ksi}) = 40 \text{ ksi} \quad \text{[OK]}$$

Practice Problem 2: 161 ft Steel I-Beam Bridge with Concrete Slab

Situation

L	span length	161 ft
LL	live load model	HL-93
S	beam spacing	13 ft
t_s	slab thickness	
	(including 0.5 in integral wearing surface)	10.5 in
	self-weight of steel girder	335 lbf/ft

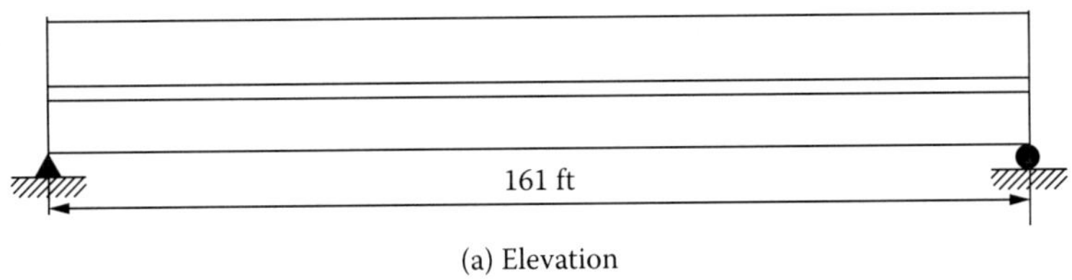

(a) Elevation

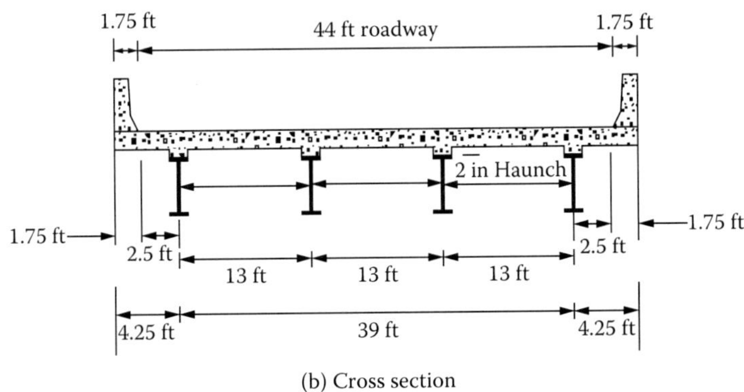

(b) Cross section

FIGURE 3.18
Steel I-beam with concrete slab.

FWS	future wearing surface	25 lbf/ft^2
	weight of stiffeners and bracings	10% of girder weight
	dead load of the curb/parapet	0.505 kips/ft
	self-weight of stay-in-place (SIP) forms	7 lbf/ft^2
f'_c	concrete strength	5.0 kips/in^2
w_c	concrete unit weight	145 lbf/ft^3

The steel girders satisfy the provisions of AASHTO Art. 6.10.1 and 6.10.2 for the cross-section properties' proportion limits. See Figure 3.18.

Solution

Steel Girder Section Review. See Figure 3.19

A	area of girder	106.0 in^2
y_b	distance from bottom fiber to centroid of section	30.88 in
y_t	distance from top fiber to centroid of section	40.12 in
I	moment of inertia about steel section centroid	99,734.0 in^4

Please see Figure 3.20.

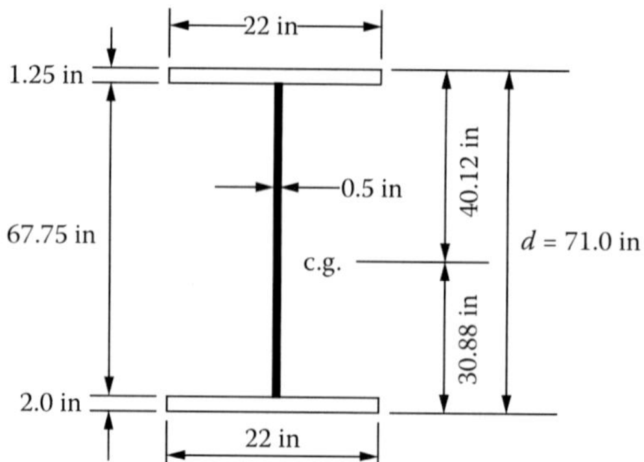

FIGURE 3.19
I-beam properties.

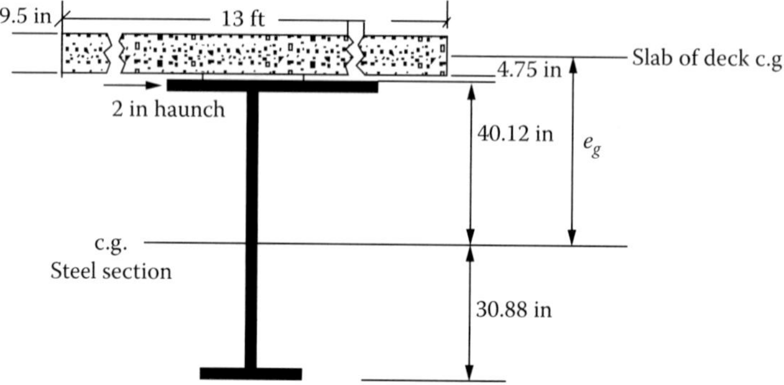

FIGURE 3.20
Cross-section properties for shears and moments due to dead loads.

Step 1: Determine Dead Load Shears and Moments for Interior and Exterior Girders

Dead Loads DC: Components and Attachments

1.1 Interior Beam

1.1.a Noncomposite Dead Loads, DC_1 The steel beam weight plus 10% for stiffeners and bracings is,

$$DC_{beam} = \left(0.335\frac{\text{kips}}{\text{ft}}\right)(1.10)$$

$$DC_{beam} = 0.369\frac{\text{kips}}{\text{ft}}$$

The slab dead load is

$$DC_{slab} = (10.5\ \text{in})\left(\frac{1\ \text{ft}}{12\ \text{in}}\right)(13\ \text{ft})\left(0.15\frac{\text{kips}}{\text{ft}^3}\right)$$

$$DC_{slab} = 1.70\frac{\text{kips}}{\text{ft}}$$

The concrete haunch dead load is

$$DC_{haunch} = (2\ \text{in})(22\ \text{in})\left(\frac{1\ \text{ft}^2}{144\ \text{in}^2}\right)\left(0.15\frac{\text{kips}}{\text{ft}}\right)$$

$$DC_{haunch} = 0.046\frac{\text{kips}}{\text{ft}}$$

The stay-in-place (SIP) forms dead load is

$$DC_{SIP} = \left(0.007\frac{\text{kips}}{\text{ft}^2}\right)(44.0\ \text{ft})\left(\frac{1}{4\ \text{girders}}\right)$$

$$DC_{SIP} = 0.077\frac{\text{kips}}{\text{ft}}$$

$$DC_1 = DC_{beam} + DC_{slab} + DC_{haunch} + DC_{SIP}$$

$$DC_1 = 0.369\frac{\text{kips}}{\text{ft}} + 1.70\frac{\text{kips}}{\text{ft}} + 0.046\frac{\text{kips}}{\text{ft}} + 0.077\frac{\text{kips}}{\text{ft}}$$

$$DC_1 = 2.19\frac{\text{kips}}{\text{ft}}$$

Shear

$$V_{DC_1} = \frac{wL}{2} = \frac{\left(2.19\frac{\text{kips}}{\text{ft}}\right)(161\ \text{ft})}{2}$$

$$V_{DC_1} = 176.3\ \text{kips}$$

Moment

$$M_{DC_1} = \frac{wL^2}{8} = \frac{\left(2.19\,\frac{kips}{ft}\right)(161\text{ ft})^2}{8}$$

$$M_{DC_1} = 7{,}095.9\text{ kip-ft}$$

1.1.b Composite Dead Loads Due to Curb/Parapet, DC_2

$$DC_2 = \left(0.505\,\frac{kips}{ft}\right)(2)\left(\frac{1}{4\text{ girders}}\right)$$

$$DC_2 = 0.253\,\frac{kips}{ft}$$

Shear

$$V_{DC_2} = \frac{wL}{2} = \frac{\left(0.253\,\frac{kips}{ft}\right)(161\text{ ft})}{2}$$

$$V_{DC_2} = 20.4\text{ kips}$$

Moment

$$M_{DC_2} = \frac{wL^2}{8} = \frac{\left(0.253\,\frac{kips}{ft}\right)(161\text{ ft})^2}{8}$$

$$M_{DC_2} = 819.8\text{ kip-ft}$$

1.1.c Total Dead Load Effects Due to $DC = DC_1 + DC_2$

Shear

$$V_{DC} = V_{DC_1} + V_{DC_2}$$

$$V_{DC} = 176.3\text{ kips} + 20.4\text{ kips}$$

$$V_{DC} = 196.7\text{ kips}$$

Moment

$$M_{DC} = M_{DC_1} + M_{DC_2}$$

$$M_{DC} = 7.095.9 \text{ kip-ft} + 819.8 \text{ kip-ft}$$

$$M_{DC} = 7{,}915.7 \text{ kip-ft}$$

1.1.d Dead Loads, DW; Wearing Surfaces

For future wearing surface of 25 lbf/ft^2

$$DW = \left(25\frac{\text{lbf}}{\text{ft}^2}\right)\left(\frac{1 \text{ kip}}{1000 \text{ lbf}}\right)(44 \text{ ft})\left(\frac{1}{4 \text{ girders}}\right)$$

$$DW = 0.275\frac{\text{kips}}{\text{ft}}$$

Shear

$$V_{DW} = \frac{wL}{2} = \frac{\left(0.275\frac{\text{kips}}{\text{ft}}\right)(161 \text{ ft})}{2}$$

$$V_{DW} = 22.14 \text{ kips}$$

Moment

$$M_{DW} = \frac{wL^2}{8} = \frac{\left(0.275\frac{\text{kips}}{\text{ft}}\right)(161 \text{ ft})^2}{8}$$

$$M_{DW} = 891.0 \text{ kip-ft}$$

Please see Table 3.3 for summary in interior beams.

1.2 Exterior Beams

The exterior beam has a slab overhang of 4.25 ft. Therefore, the interior beam dead loads control for design.

TABLE 3.3
Summary of Dead Load Shears and Moments in Interior Beams

	DC			
	DC_1	DC_2	$DC_1 + DC_2$	DW
Shear, kips	176.3	20.4	196.7	22.1
Moment, kip-ft	7095.9	819.8	7915.7	891.0

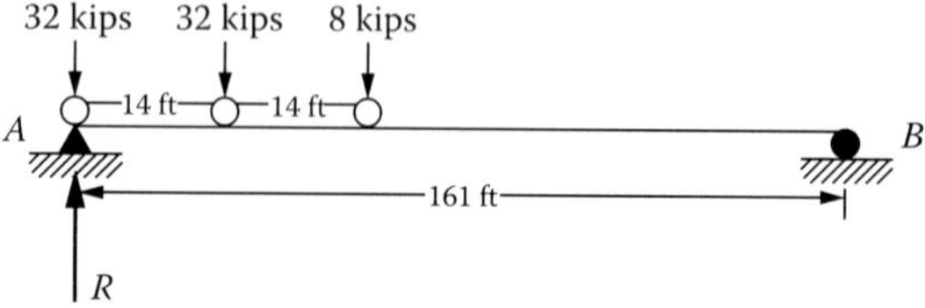

FIGURE 3.21
Design truck load (HS-20) position for maximum shear.

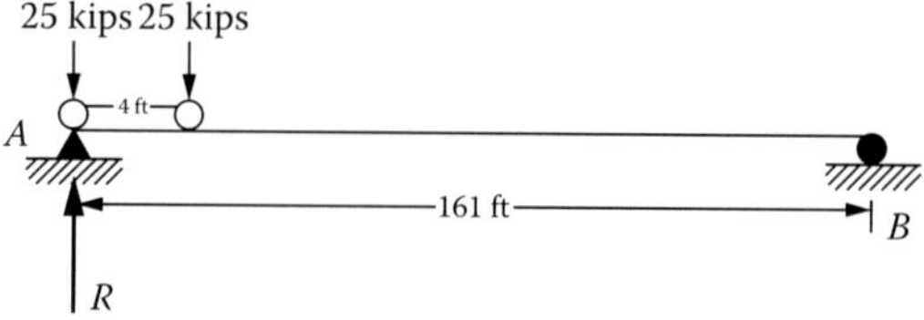

FIGURE 3.22
Design tandem load position for maximum shear.

Step 2: Determine Live Load Shears and Moments Using the AASHTO HL-93 Load.

2.1 Determine Live Load Shears Due to HL-93 Loads

Please see Figure 3.21.

$$\Sigma M_B = 0$$

$$(32\text{ kips})(161\text{ ft}) + (32\text{ kips})(161\text{ ft} - 14\text{ ft}) + (8\text{ kips})(161\text{ ft} - 28\text{ ft}) - R(161\text{ ft}) = 0$$

$$R = 67.8\text{ kips}$$

$$V_{truck} = 67.8\text{ kips [controls]}$$

Please see Figure 3.22.

$$\Sigma M_B = 0$$

$$(25 \text{ kips})(161 \text{ ft}) + (25 \text{ kips})(161 \text{ ft} - 4 \text{ ft}) - R(161 \text{ ft}) = 0$$

$$R = 49.4 \text{ kips}$$

$$V_{tandem} = 49.4 \text{ kips}$$

Please see Figure 3.23.

$$\Sigma M_B = 0$$

$$(0.640 \text{ kips/ft})(161 \text{ ft})(1/2) - R(161 \text{ ft}) = 0$$

$$R = 67.8 \text{ kips}$$

$$V_{lane} = 51.5 \text{ kips}$$

2.2 Determine Live Load Moments Due to HL-93 Loads

Please see Figure 3.24 for design truck moment.

$$\Sigma M_B = 0$$

$$(32 \text{ kips})(80.5 \text{ ft} + 14 \text{ ft}) + (32 \text{ kips})(80.5 \text{ ft}) + (8 \text{ kips})$$
$$(80.5 \text{ ft} - 14 \text{ ft}) - R(161 \text{ ft}) = 0$$

$$R = 38.09 \text{ kips}$$

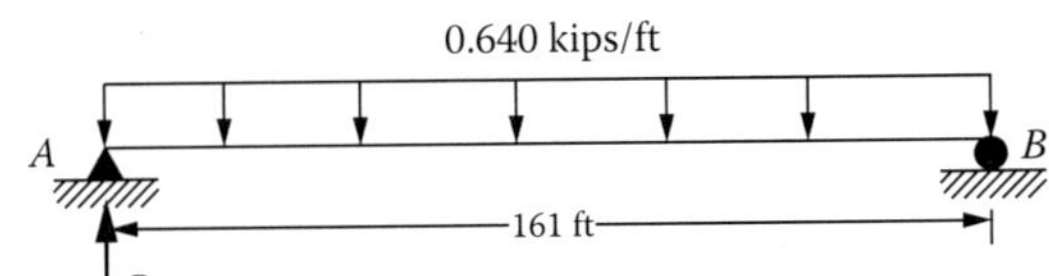

FIGURE 3.23
Design lane load position for maximum shear.

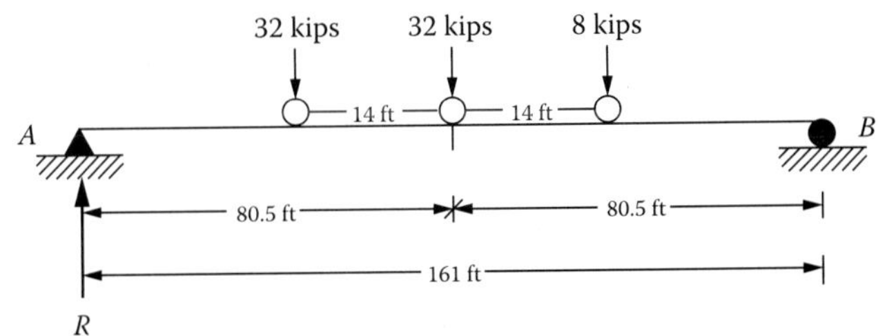

FIGURE 3.24
Design truck load (HS-20) position for maximum moment.

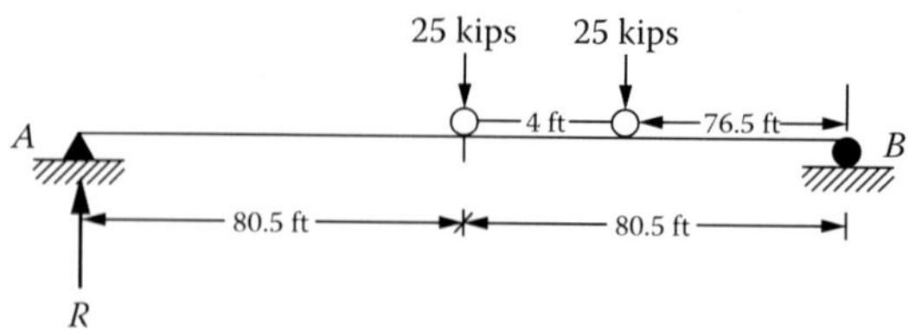

FIGURE 3.25
Design tandem load position for maximum moment.

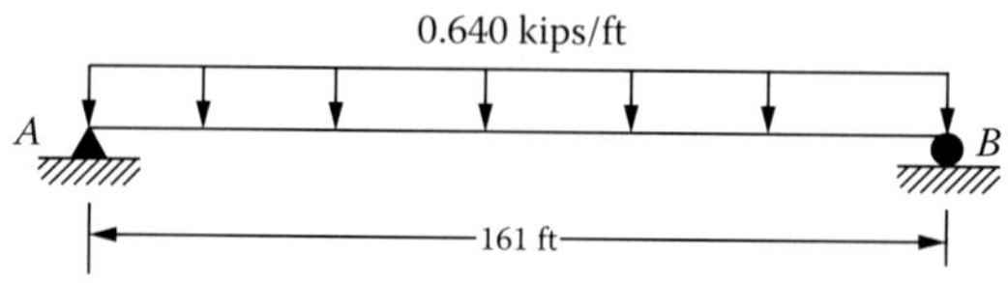

FIGURE 3.26
Design lane load position for maximum moment.

$$M_{truck} = (38.09 \text{ kips})(80.5 \text{ ft}) - (32 \text{ kips})(14 \text{ ft})$$

$$M_{truck} = 2618.2 \text{ kip-ft [controls]}$$

Please see Figure 3.25 for design tandem load moment.

$$\Sigma M_B = 0$$

$$(25 \text{ kips})(80.5 \text{ ft}) + (25 \text{ kips})(80.5 \text{ ft} - 4 \text{ ft}) - R(161 \text{ ft}) = 0$$

$$R = 24.38 \text{ kips}$$

$$M_{tandem} = (24.38 \text{ kips})(80.5 \text{ ft})$$

$$M_{tandem} = 1962.6 \text{ kip-ft}$$

Please see Figure 3.26 for design lane moment.

$$M_{lane} = \frac{wL^2}{8} = \frac{\left(0.640 \frac{\text{kips}}{\text{ft}}\right)(161 \text{ ft})^2}{8}$$

$$M_{lane} = 2073.7 \text{ kip-ft}$$

Please see Table 3.4, summary of live load shears and moments.

TABLE 3.4

Summary of Live Load Shears and Moments

	Truck Load Effect	Tandem Load Effect	Lane Load Effect
Shear, kips	67.8	49.4	51.5
Moment, kip-ft	2618.2	1962.6	2073.7

Step 3: Determine Live Load Distribution Factors

A Art. 4.6.2.2.2; Tbl. 4.6.2.2.1-1

3.1 For Moment in Interior Beam, DFM

A Art. 4.6.2.2.2b

Longitudinal stiffness parameter K_g is,

A Eq. 4.6.2.2.1-1

$$K_g = n\left(I + A e_g^2\right)$$

where:

$$n = \frac{E_s}{E_c}$$

E_s = modulus of elasticity of steel = 29 x 10^3 kips/in^2
E_c = modulus of elasticity of deck concrete material

A Comm., Eq. 5.4.2.4-1

$E_c = 1820\sqrt{f_c'}$ for w_c = 0.145 kips/ft^3

$$= 1820\sqrt{5\frac{\text{kips}}{\text{in}^2}}$$

$$= 4069.6\frac{\text{kips}}{\text{in}^2}$$

e_g = distance between the centers of gravity of the beam and deck
e_g = 40.12 in + 2 in + 4.75 in
e_g = 46.87 in

$$n = \frac{29 \times 10^3}{4.0696 \times 10^3} = 7.13$$

Use n = 7.0.

A = area of basic beam section
A = 106.0 in^2 (given)
I = moment of inertia of basic beam section
I = 99,734.0 in^4 (given)
K_g = 7.0 [99734.0 in^4 + (106 in^2)(46.87 in)^2]
= 2.33 x 10^6 in^4

Distribution of live loads per lane for moment in interior beams with cross-section type (a).

A Tbl. 4.6.2.2.1-1, 4.6.2.2.2b-1 or Appendix A

One design lane loaded:

$$(m)(DFM_{si}) = DFM_{si} = 0.06 + \left(\frac{S}{14}\right)^{0.4}\left(\frac{S}{L}\right)^{0.3}\left(\frac{K_g}{12\,Lt_s{}^3}\right)^{0.1}$$

where:

m = multiple presence factors have been included in the approximate equations for distribution factors. They are applied where the lever rule is used.

A Art. 3.6.1.1.2; Tbl. 3.6.1.1.2-1

si = single lane loaded, interior girder
DFM = moment distribution factor
S = spacing of beams (ft)
L = span length of beam (ft)
t_s = depth of concrete slab (in)

$$DFM_{si} = 0.06 + \left(\frac{13\text{ ft}}{14}\right)^{0.4}\left(\frac{13\text{ ft}}{161\text{ ft}}\right)^{0.3}\left(\frac{2.33\times10^6\text{ in}^4}{12(161\text{ ft})(9.5\text{ in})^3}\right)^{0.1}$$

$$= 0.532$$

Two or more lanes loaded:

$$DFM_{mi} = 0.075 + \left(\frac{S}{9.5}\right)^{0.6}\left(\frac{S}{L}\right)^{0.2}\left(\frac{K_g}{12\,Lt_s^3}\right)^{0.1}$$

where:
mi = multiple lanes loaded, interior girder

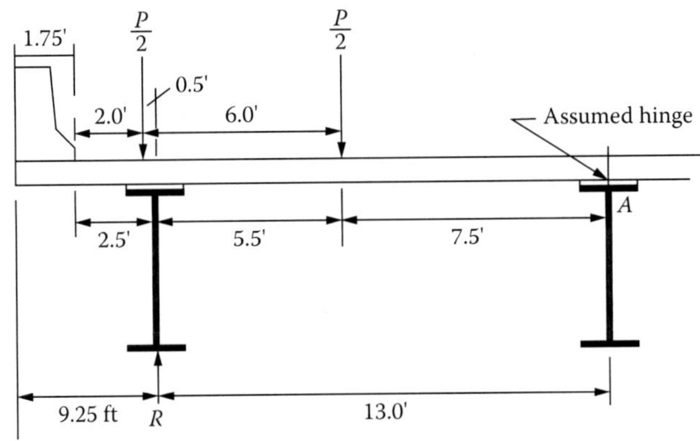

FIGURE 3.27
Lever rule for the distribution factor for moments for exterior girder.

$$\mathrm{DFM_{mi}} = 0.075 + \left(\frac{13\text{ ft}}{9.5}\right)^{0.6}\left(\frac{13\text{ ft}}{161\text{ ft}}\right)^{0.2}\left(\frac{2.33\times10^{6}\text{ in}^4}{12(161\text{ ft})(9.5\text{ in})^3}\right)^{0.1}$$

$$= 0.83$$

3.2 Distribution of Live Load per Lane for Moment in Exterior Beams with Cross-Section Type (a)

A Tbl. 4.6.2.2.2d-1 or Appendix B

One design lane loaded:

Use lever rule. See Figure 3.27.

A Fig. C4.6.2.2.1-1

$$\Sigma M_A = 0$$

$$(P/2)(13.5\text{ ft}/2) + (P/2)(7.5\text{ ft}) - R(13\text{ ft}) = 0$$

$$R = 0.808\text{ P}$$

$$\mathrm{DFM_{se}} = (m)(0.808)$$

$$R = (1.2)(0.808) = 0.970$$

where:
se = single lane loaded, exterior girder
m = multiple presence factor for one lane loaded = 1.2

Two or More Design Lanes Loaded:

$g = (e)(g_{interior})$ or
$DFM_{me} = (e)DFM_{mi}$

$$e = 0.77 + \frac{d_e}{9.1}$$

where:
g = DFM = distribution factors for moment
$g_{interior}$ = DFM_{mi} = distribution factor for interior girder
= 0.83
d_e = distance from the centerline of the exterior beam web to the interior edge of curb (ft)
= 2.5 ft

$$e = 0.77 + \frac{2.5\ \text{ft}}{9.1} = 1.045\ \text{ft}$$

$$DFM_{me} = (1.045\ \text{ft})(0.83)$$

$$= 0.867 \text{ or}$$

$$DFM_{me} = g = 0.867$$

3.3 Distribution of Live Load per Lane for Shear in Interior Beams.

A Tbl. 4.6.2.2.3a-1 or Appendix C

One design lane loaded:

$$DFV_{si} = 0.36 + \frac{S}{25.0}$$

$$= 0.36 + \frac{13\ \text{ft}}{25.0}$$

$$= 0.88$$

where:
DFV = shear distribution factor

Two or more lanes loaded:

$$DFV_{mi} = 0.2 + \frac{S}{12} - \left(\frac{S}{35}\right)^{2.0}$$

$$= \left(0.2 + \frac{13\text{ ft}}{12} - \left(\frac{13\text{ ft}}{35}\right)^{2.0}\right)$$

$$= 1.145$$

3.4 Distribution of Live Load per Lane for Shear in Exterior Beams

A Tbl. 4.6.2.2.3b-1 or Appendix D

One design lane loaded:

Use lever rule (same as the distribution factors for moment)

$$DFV_{se} = DFM_{se}$$

$$= 0.970$$

Two or more design lanes loaded:

$$DFV_{se} = (e)(g_{interior})$$

$$g_{interior} = DFV_{mi}$$

$$DFV_{me} = (e)(DFV_{mi})$$

where:

$$e = 0.6 + \frac{d_e}{10}$$

$$e = 0.6 + \frac{2.5\text{ ft}}{10}$$

$$e = 0.85$$

$$DFV_{me} = (0.85)(1.145)$$

$$= 0.973$$

TABLE 3.5

Summary of Live Load Distribution Factors

		Moment		Shear	
	Distribution Factor Equation	Interior Beam	Exterior Beam	Interior Beam	Exterior Beam
One Lane Loaded	Approximate	0.532	–	0.88	—
	Lever Rule	—	0.970	—	0.970
Two or More Lanes Loaded	Approximate	0.830	0.867	1.145	—
	Lever Rule	—	—	—	0.973
Controlling Value		0.830	0.970	1.145	0.973

Please see Table 3.5.

Step 4: Unfactored Distributed Live Load per Beam with Impact

Dynamic load allowance, IM, is equal to

33% for design truck or tandem

0% for design lane

15% for Fatigue and Fracture Limit State

A Art. 3.6.2.1

4.1 Shear

$$V_{LL+IM} = DFV\left[\left(V_{truck} \text{ or } V_{tandem}\right)\left(1+IM\right)+V_{lane}\right]$$

4.1.a Interior Beam

$$V_{LL+IM} = (1.145)\left[(67.8 \text{ kips})(1.33)+51.5 \text{ kips}\right]$$

$$V_{LL+IM} = 162.2 \text{ kips}$$

4.1.b Exterior Beam

$$V_{LL+IM} = (0.973)\left[(67.8 \text{ kips})(1.33)+51.5 \text{ kips}\right]$$

$$V_{LL+IM} = 137.8 \text{ kips}$$

TABLE 3.6

Summary of Unfactored Distributed Live Load Effects per Beam

Beams	V_{LL+IM}, kips	M_{LL+IM}, kip-ft
Interior	162.2	4611.4
Exterior	137.8	5389.2

4.2 Moment

$$M_{LL+IM} = DFM\left[\left(M_{truck} \text{ or } M_{tandem}\right)\left(1+IM\right)+M_{lane}\right]$$

4.2.a Interior Beam

$$M_{LL+IM} = 0.83\left[\left(2618.2 \text{ kip-ft}\right)\left(1.33\right)+2073.7 \text{ kip-ft}\right]$$

$$M_{LL+IM} = 4611.4 \text{ kip-ft}$$

4.2.b Exterior Beam

$$M_{LL+IM} = 0.97\left[\left(2618.2 \text{ kip-ft}\right)\left(1.33\right)+2073.7 \text{ kip-ft}\right]$$

$$M_{LL+IM} = 5389.2 \text{ kip-ft}$$

Please see Table 3.6.

Step 5: Factored Design Loads for Limit State Strength I

A Tbl. 3.4.1-1, 3.4.1-2

5.1 Shear, V_u

$$V_u = 1.25V_{DC} + 1.5V_{DW} + 1.75V_{LL+IM}$$

5.1.a Interior Beam

$$V_u = 1.25\left(196.7 \text{ kips}\right)+1.5\left(22.1 \text{ kips}\right)+1.75\left(162.2 \text{ kips}\right)$$

$$V_u = 562.9 \text{ kips}$$

5.1.b Exterior Beam

$$V_u = 1.25(196.7 \text{ kips}) + 1.5(22.1 \text{ kips}) + 1.75(137.8 \text{ kips})$$

$$V_u = 520.2 \text{ kips}$$

5.2 Moment, M_u

$$M_u = 1.25M_{DC} + 1.5M_{DW} + 1.75M_{LL+IM}$$

5.2.a Interior Beam

$$M_u = 1.25(7915.7 \text{ kip-ft}) + 1.5(891.0 \text{ kip-ft}) + 1.75(4611.4 \text{ kip-ft})$$

$$M_u = 19301.0 \text{ kip-ft}$$

5.2.b Exterior Beam

$$M_u = 1.25(7915.7 \text{ kip-ft}) + 1.5(891.0 \text{ kip-ft}) + 1.75(5389.2 \text{ kip-ft})$$

$$M_u = 20662.2 \text{ kip-ft}$$

Step 6: Fatigue II Limit State for Finite Load-Induced Fatigue Life

A Arts. 3.6.1.4, 3.6.1.1.2, 3.6.2.1

6.1 Fatigue Load

Fatigue load shall be one design truck or axles, but with a constant spacing of 30 ft between the 32 kip axles. Please see Figure 3.28.

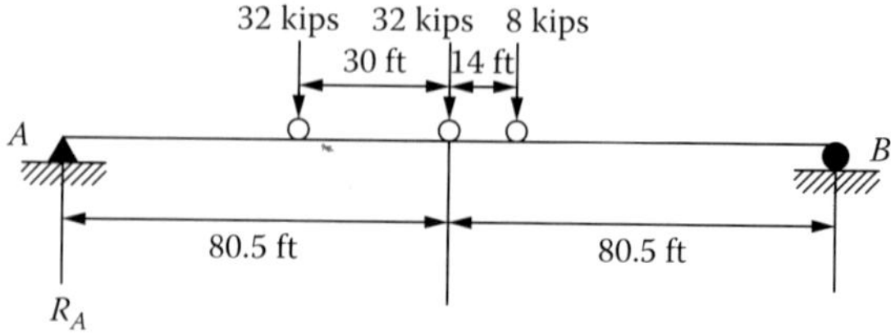

FIGURE 3.28
Fatigue load position for maximum moment at midspan.

$$\Sigma M_B = 0$$

$$(32 \text{ kips})(80.5 \text{ ft} + 30 \text{ ft}) + (32 \text{ kips})(80.5 \text{ ft}) + (8 \text{ kips})(80.5 \text{ ft} - 14 \text{ ft}) - R_A(161.0 \text{ ft}) = 0$$

$$R_A = 41.27 \text{ kips}$$

The maximum fatigue load moment, M_{fat}, is,

$$M_{fat} = (41.27 \text{ kips})(80.5 \text{ ft}) - (32 \text{ kips})(30.0 \text{ ft})$$

$$M_{fat} = 2362.0 \text{ kip-ft}$$

The multiple presence factors have been included in the approximate equations for distribution factors in AASHTO [4.6.2.2 and 4.6.2.3], both for single and multiple lanes loaded. Where the lever rule (a sketch is required to determine load distributions) is used, the multiple presence factors must be included. Therefore, for fatigue investigations in which fatigue truck is placed in a single lane, the factor 1.2 which has been included in the approximate equations should be removed.

A Comm. 3.6.1.1.2

6.2 Fatigue Live Load Distribution Factors per Lane for Moment

The dynamic allowance for Fatigue Limit State, IM, is 15%.

A Tbl. 3.6.2.1-1

Determine distribution factors for moments, DFM_{fat},

6.2.a For Interior Beam

$$DFM^{I}_{fat} = (DFM_{si})\left(\frac{1}{m}\right)$$

$$DFM^{I}_{fat} = (0.532)\left(\frac{1}{1.2}\right)$$

$$DFM^{I}_{fat} = 0.44$$

6.2.b For Exterior Beam

$$DFM^{E}_{fat} = (DFM_{se})\left(\frac{1}{m}\right)$$

$$DFM^{E}_{fat} = (0.97)\left(\frac{1}{1.2}\right)$$

$$DFM^{E}_{fat} = 0.81$$

6.3 Unfactored Distributed Fatigue Live Load Moment per Beam with Impact

$$M_{fat+IM} = (DFM_{fat})(M_{fat})(1+IM)$$

6.3.a Interior Beam

$$M^{I}_{fat+IM} = (DFM^{I}_{fat})(M_{fat})(1+IM)$$

$$M^{I}_{fat+IM} = (0.44)(2362.0\text{ kip-ft})(1+0.15)$$

$$M^{I}_{fat+IM} = 1195.2\text{ kip-ft}$$

6.3.b Exterior Beam

$$M^{E}_{fat+IM} = (DFM^{E}_{fat})(M_{fat})(1+IM)$$

$$M^{E}_{fat+IM} = (0.81)(2362.0\text{ kip-ft})(1+0.15)$$

$$M^{I}_{fat+IM} = 2200.2\text{ kip-ft}$$

6.4 Factored Fatigue II Design Live Load Moment per Beam, M_Q

A Tbl. 3.4.1-1

$$Q = 0.75\ M_{LL+IM}$$

6.4.a Interior Beam

$$M^{I}_{Q,fat} = 0.75(M^{I}_{fat+IM})$$

$$M^{I}_{Q,fat} = 0.75(1195.2\text{ kip-ft})$$

$$M^{I}_{Q,fat} = 896.4\text{ kip-ft}$$

6.4.b Exterior Beam

$$M^{E}_{Q,fat} = 0.75\left(M^{E}_{fat+IM}\right)$$

$$M^{I}_{Q,fat} = 0.75\left(2200.2 \text{ kip-ft}\right)$$

$$M^{I}_{Q,fat} = 1650.2 \text{ kip-ft}$$

Practice Problem 3: Interior Prestressed Concrete I-Beam

Situation

The following design specifications apply to a bridge in western Pennsylvania.

L	bridge span	80 ft
f'_{ci}	compressive strength of concrete at time of initial prestress	5.5 ksi
f'_{cg}	compressive strength of concrete at 28 days for prestressed I-beams	6.5 ksi
	number of roadway prestressed 24 × 54 in I-beams	6 beams
S	beam spacing	8 ft
t_s	roadway slab thickness	7.5 in
	integral wearing surface of slab	0.50 in
	clear roadway width	44 ft 6 in
f'_{cs}	compressive strength of roadway slab concrete at 28 days	4.5 ksi
w_{FWS}	future wearing surface dead load	0.030 kips/ft^2
w_s	$= w_{C\&P}$ curb and parapet dead load	0.506 kips/ft
A_{ps}	area of prestressing steel (0.5 in diameter low-relaxation strand; seven wire)	0.153 in^2
E_p	modulus of elasticity of prestressing steel	28,500 ksi
f_{pu}	ultimate stress of prestressing steel (stress relieved)	270 ksi

A composite deck and a standard curb and parapet are used. Assume no haunch for composite section properties and 150 lbf/ft^3 concrete for all components. The bridge cross section is shown. See Figures 3.29, 3.30, and 3.31.

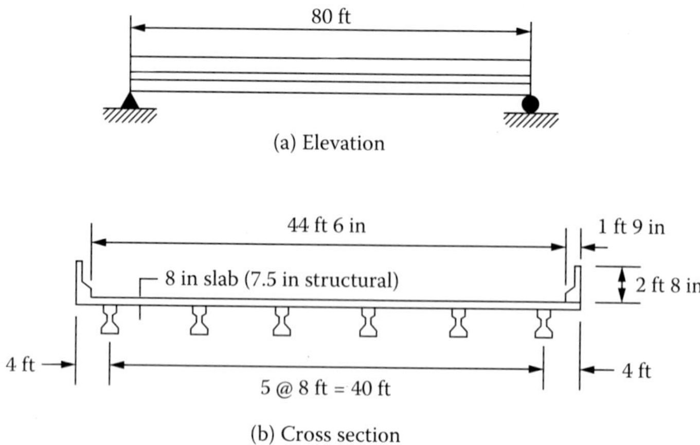

FIGURE 3.29
Prestressed concrete I-beam.

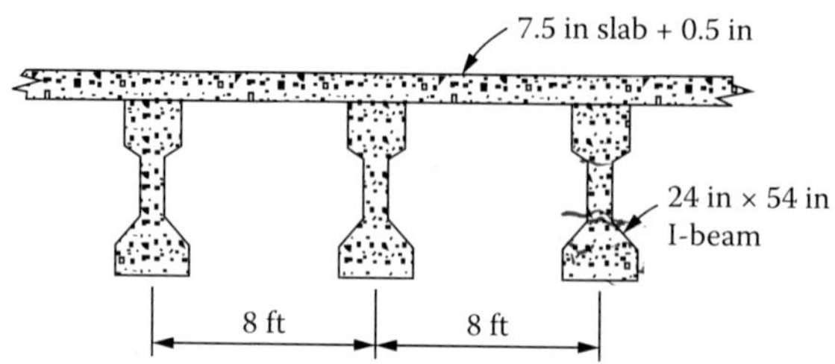

FIGURE 3.30
Deck and I-beam.

Requirement

Review the interior bonded prestressed concrete I-beam for the Load Combination Limit State Strength I.

where:
- Q_i = force effect
- η = load modifier
- γ = load factor
- $\eta_i = \eta_D \eta_R \eta_I = 1.0$

Solution

Strength I Limit State: Q = 1.0(1.25 DC + 1.50 DW + 1.75(LL + IM)). Please see Table 3.7.

A Tbl. 3.4.1-1, A Art. 1.3.3, 1.3.4, 1.3.5, 1.3.2.1

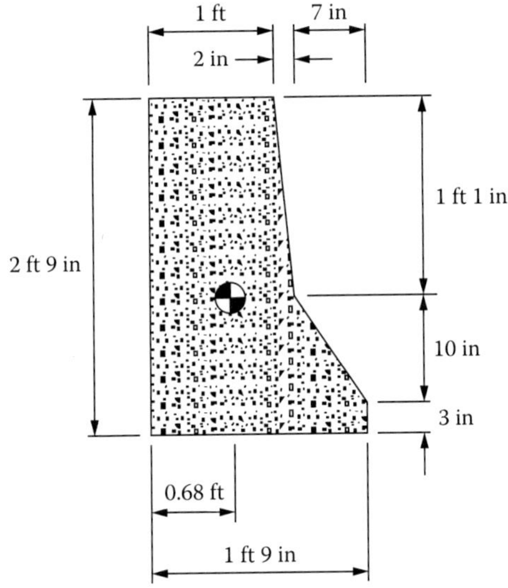

FIGURE 3.31
Curb and parapet.

TABLE 3.7
Load Modifiers

	Strength	Service	Fatigue
Ductility, η_d	1.0	1.0	1.0
Redundancy, η_r	1.0	1.0	1.0
Importance, η_i	1.0	N/A	N/A
$\eta = \eta_D \eta_R \eta_I \geq 0.95$	1.0	1.0	1.0

Step 1: Determine the Cross-Sectional Properties for a Typical Interior Beam

The properties of the basic beam section are given as follows:

A_g	area of basic beam	816 in^2
y_b	center of gravity of the basic beam from the bottom of the basic beam	25.31 in
y_t	distance from the center of gravity of the basic beam to the top of the basic beam	28.69 in
h	basic beam depth	54.0 in
I_g	moment of inertia of the basic beam section about centroidal axis, neglecting reinforcement	255,194 in^4

The section moduli for the extreme fiber of the noncomposite section are

$$S_{ncb} = \frac{I_g}{y_b} = \frac{255{,}194\ \text{in}^4}{25.31\ \text{in}} = 10{,}083\ \text{in}^3$$

$$S_{nct} = \frac{I_g}{h - y_b} = \frac{255{,}194\ \text{in}^4}{54.0\ \text{in} - 25.31\ \text{in}} = 8895\ \text{in}^3$$

The unit weight of concrete, w_c, is 0.150 kips/ft^3.

A Tbl. 3.5.1-1

The modulus of elasticity of concrete at transfer is

A Eq. 5.4.2.4-1

$$E_{ci} = 33{,}000\ w_c^{1.5}\sqrt{f'_{ci}} = (33{,}000)\left(0.150\frac{\text{kips}}{\text{ft}^3}\right)^{1.5}\sqrt{5.5\ \text{ksi}} = 4496\ \text{ksi}$$

The modulus of elasticity for the prestressed I-beam concrete at 28 days is

$$E_{cg} = 33{,}000\ w_c^{1.5}\sqrt{f'_{cg}} = (33{,}000)\left(0.150\frac{\text{kips}}{\text{ft}^3}\right)^{1.5}\sqrt{6.5\ \text{ksi}} = 4888\ \text{ksi}$$

The modulus of elasticity of the roadway slab concrete at 28 days is

$$E_{cs} = 33{,}000\ w_c^{1.5}\sqrt{f'_{cs}} = (33{,}000)\left(0.150\frac{\text{kips}}{\text{ft}^3}\right)^{1.5}\sqrt{4.5\ \text{ksi}} = 4067\ \text{ksi}$$

Please see Figure 3.32.

b_w	web width	8 in
b_{top}	top flange width	18 in

The effective flange width for interior beams is

A Art. 4.6.2.6.1

$$b_i = S = 8\ \text{ft} = 96\ \text{in}$$

The modular ratio of the slab and beam is

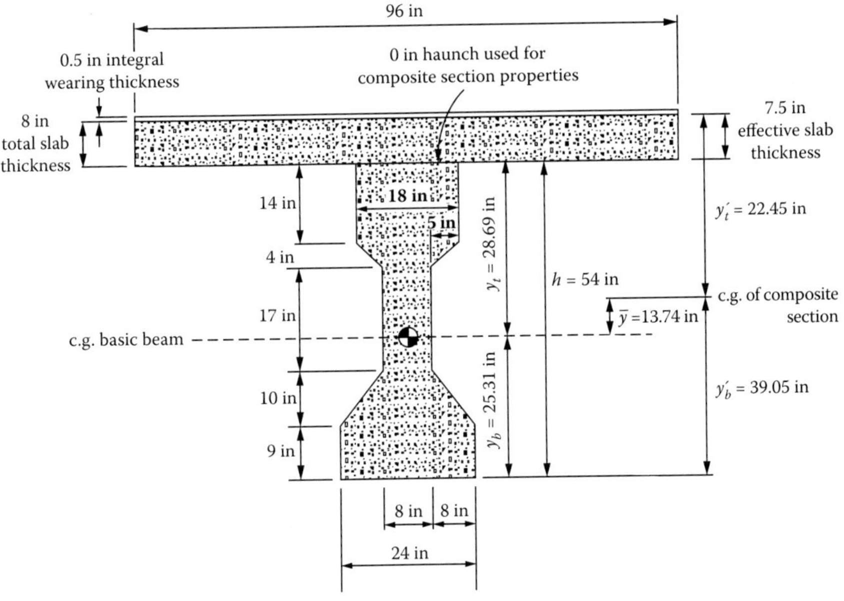

FIGURE 3.32
Composite section.

$$n = \frac{E_{cs}}{E_{cg}} = \frac{4067 \text{ ksi}}{4888 \text{ ksi}} = 0.832$$

Therefore, the transformed flange width is

$$(b_i)(n) = (96 \text{ in})(0.832) = 79.9 \text{ in}$$

Transformed flange area is

$$(b_i n)(t_s) = (79.9 \text{ in})(7.5 \text{ in}) = 599.25 \text{ in } (600.0 \text{ in}^2)$$

The area of transformed gross composite section, A_{gc}, is

$$A_{gc} = A_g + (b_i)(n)(t_s)$$

$$= 816 \text{ in}^2 + 600.0 \text{ in}^2$$

$$= 1416 \text{ in}^2$$

Use the transformed flange area to compute composite section properties by summing moments of areas about the centroid of the basic beam section.

$$A_{gc}(\bar{y}) = (1416\ \text{in}^2)\bar{y} = (600\ \text{in}^2)\left(y_t + \frac{7.5\ \text{in}}{2}\right)$$

$$= (600\ \text{in}^2)(28.69\ \text{in} + 3.75\ \text{in})$$

$$= 19{,}464\ \text{in}^3$$

$\bar{y}$ = 13.74 in from the centroid of the basic beam.

Thus, for the composite section,

$$y_b' = y_b + \bar{y} = 25.31\ \text{in} + 13.74\ \text{in} = 39.05\ \text{in}$$

$$y_t' = (h + t_s) - y_b' = (54.0\ \text{in} + 7.5\ \text{in}) - 39.05\ \text{in}$$

$$= 22.45\ \text{in}$$

The composite moment of inertia about the centroid of the composite section is

$$I_c = I_g + A_g\bar{y}^2 + \frac{b_i n t_s^3}{12} + (600\ \text{in}^2)\left(y_t' - \frac{t_s}{2}\right)^2$$

$$= 255{,}194\ \text{in}^4 + (816\ \text{in}^2)(13.74\ \text{in})^2 + \frac{(96\ \text{in})(0.832)(7.5\ \text{in})^3}{12}$$

$$+ (600\ \text{in}^2)\left(22.45\ \text{in} - \frac{7.5\ \text{in}}{2}\right)^2$$

$$= 621{,}867\ \text{in}^4$$

The section modulus of the composite section for the bottom extreme fiber is

$$S_{cb} = \frac{I_c}{y_b'} = \frac{621{,}867\ \text{in}^4}{39.05\ \text{in}} = 15{,}925\ \text{in}^3$$

The section modulus of the composite section for the top extreme fiber is

$$S_{ct} = \frac{I_c}{y_t'} = \frac{621{,}867\ \text{in}^4}{22.45\ \text{in}} = 27{,}700\ \text{in}^3$$

Step 2: Perform the Dead Load Analysis

The beam weight is

$$w = \left(816\ \text{in}^2\right)\left(\frac{1\ \text{ft}}{12\ \text{in}}\right)^2\left(0.15\frac{\text{kips}}{\text{ft}^3}\right) = 0.85\ \text{kips/ft}$$

The dead weight of the slab is

$$w_D = \left(8\ \text{ft}\right)\left(0.15\frac{\text{kips}}{\text{ft}^3}\right)\left(7.5\ \text{in}\right)\left(\frac{1\ \text{ft}}{12\ \text{in}}\right) = 0.75\ \text{kips/ft}$$

$$w_{DC1} = w + w_D$$

$$= 0.85\ \text{kips/ft} + 0.75\ \text{kips/ft} = 1.60\ \text{kips/ft}$$

The superimposed dead load, w_s, consists of the parapet and curb loads, distributed equally to the 6 beams.

$$w_s = \left(\frac{\left(0.506\frac{\text{kips}}{\text{ft}}\right)(2)}{6\ \text{beams}}\right) = 0.169\ \text{kips/ft}$$

$$w_s = w_{DC2} = 0.169\ \text{kips/ft}$$

$$w_{DC} = 1.60\ \text{kips/ft} + 0.169\ \text{kips/ft} = 1.769\ \text{kips/ft}$$

The future wearing surface dead load, w_{FWS}, is assumed to be equally distributed to each girder.

$$w_{FWS} = \left(8\ \text{ft}\right)\left(0.03\frac{\text{kips}}{\text{ft}}\right) = 0.240\ \text{kips/ft}$$

$$w_{FWS} = w_{DW} = 0.240\ \text{kips/ft}$$

The maximum dead load moments and shears are

$$M_{max} = \frac{wL^2}{8} = \frac{w\left(80\ \text{ft}\right)^2}{8} = 800\ w\ \text{ft-kips}$$

$$V_{max} = \frac{wL}{2} = \frac{w\left(80\ \text{ft}\right)}{2} = 40\ w\ \text{ft-kips}$$

TABLE 3.8

Dynamic Load Allowance, IM

Component	IM (%)
Deck joints, all limit states	75
Fatigue and fracture limit state	15
All other limit states	33

Step 3: Calculate the Live Load Force Effects for Moment and Shear.

Where w is the clear roadway width between curbs and/or barriers. The number of lanes is

A Art. 3.6.1.1.1

$$N_L = \frac{w}{12} = \frac{48\ \text{ft}}{12} = 4\ \ (4\ \text{lanes})$$

A Tbl. 3.6.2.1-1

Please see Table 3.8.

The distance between the centers of gravity of the basic beam and the deck is

$$e_g = y_t + \frac{t_s}{2} = 28.69\ \text{in} + \frac{7.5\ \text{in}}{2} = 32.44\ \text{in}$$

The area of the basic beam, A, is 816 in^2.

The moment of inertia of the basic beam, I_g, is 255,194 in^4.

The modular ratio between the beam and slab is

$$n = \frac{E_{cg}}{E_{cs}} = \frac{4888\ \text{ksi}}{4067\ \text{ksi}} = 1.2$$

The longitudinal stiffness parameter is

A Eq. 4.6.2.2.1-1

$$K_g = n(I_g + Ae_g^2) = (1.2)(255{,}194\ \text{in}^4 + (816\ \text{in}^2)(32.44\ \text{in})^2)$$

$$= 1{,}336{,}697.4$$

Distribution factor for moments per lane in interior beams with concrete deck,

A Art. 4.2.2.2b

$$\frac{K_g}{12\,Lt_s^3} = \frac{1{,}336{,}697.4}{12(80\text{ ft})(7.5\text{ in})^3} = 3.30$$

Common deck type (j)

A Tbl. 4.6.2.2.1-1

The multiple presence factors apply to the lever rule case only. They have been included in the approximate equations for distribution factors in Arts. 4.6.2.2 and 4.6.2.3 for both single and multiple lanes loaded.

A Comm. 3.6.1.1.2

The distribution factor for moment for interior beams per lane with one design lane loaded is

A Tbl. 4.6.2.2.2b-1 or Appendix A

$$\text{DFM}_{si} = 0.06 + \left(\frac{S}{14}\right)^{0.4}\left(\frac{S}{L}\right)^{0.3}\left(\frac{K_g}{12\,Lt_s^3}\right)^{0.1}$$

$$= 0.06 + \left(\frac{8\text{ ft}}{14}\right)^{0.4}\left(\frac{8\text{ ft}}{80\text{ ft}}\right)^{0.3}(3.30)^{0.1}$$

$$= 0.51\text{ lanes per beam}$$

The distribution factor for moment for interior beams with two or more design lanes loaded is

$$\text{DFM}_{mi} = 0.075 + \left(\frac{S}{9.5}\right)^{0.6}\left(\frac{S}{L}\right)^{0.2}\left(\frac{K_g}{12\,Lt_s^3}\right)^{0.1}$$

$$= 0.075 + \left(\frac{8\text{ ft}}{9.5}\right)^{0.6}\left(\frac{8\text{ ft}}{80\text{ ft}}\right)^{0.2}(3.30)^{0.1}$$

$$= 0.716\text{ lanes per beam [controls]}$$

The distribution factor for shear for interior beams (one design lane loaded) is

A Art. 4.6.2.2.3; Tbl. 4.6.2.2.3a-1 or Appendix C

$$DFV_{si} = 0.36 + \left(\frac{S}{25}\right) = 0.36 + \left(\frac{8.0 \text{ ft}}{25}\right)$$

$$= 0.68 \text{ lanes per beam}$$

The distribution factor for shear for interior beams (two or more design lanes loaded) is

$$DFV_{mi} = 0.2 + \left(\frac{S}{12}\right) - \left(\frac{S}{35}\right)^2 = 0.2 + \left(\frac{8.0 \text{ ft}}{12}\right) - \left(\frac{8 \text{ ft}}{35}\right)^2$$

$$= 0.814 \text{ lanes per beam [controls]}$$

Step 4: Determine the Maximum Live Load Moments and Shears

Approximate the maximum bending moment at midspan due to the HL-93 loading. Please see Figures 3.33 through 3.36.

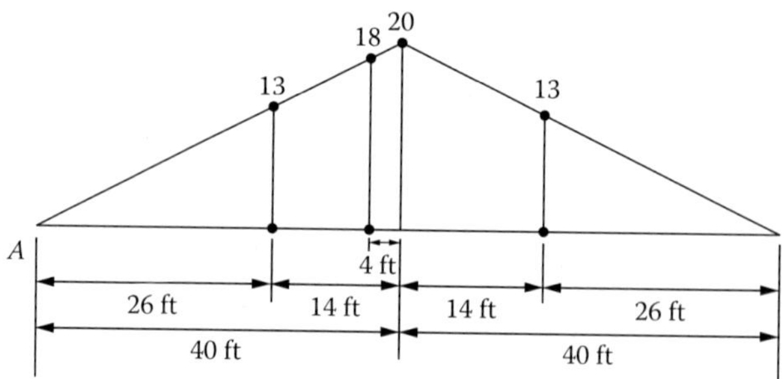

FIGURE 3.33
Influence line diagram for maximum moment at midspan.

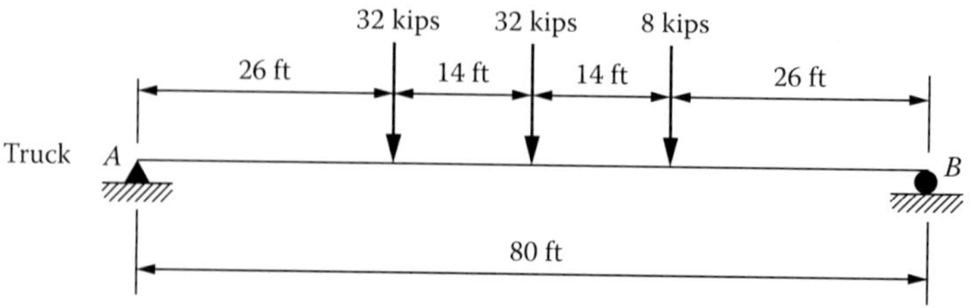

FIGURE 3.34
Design truck (HS-20) position for moment at midspan.

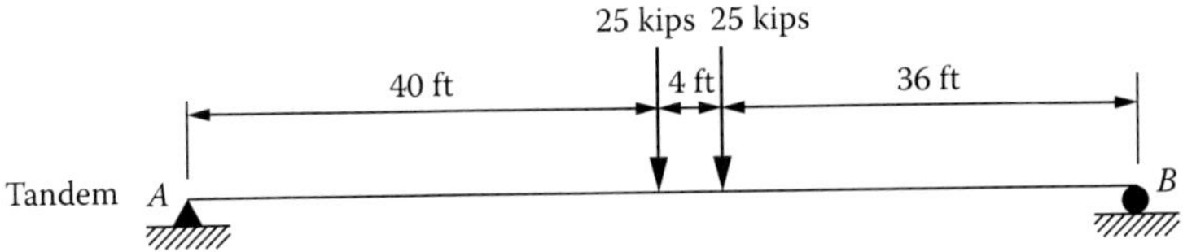

FIGURE 3.35
Design tandem load position for moment at midspan.

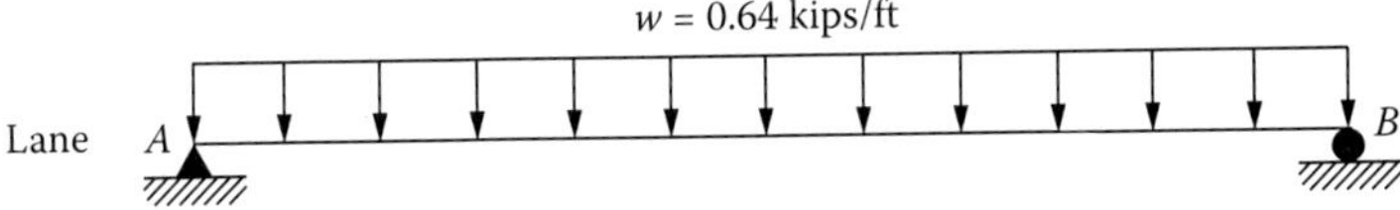

FIGURE 3.36
Design lane load for moment at midspan.

Using the influence line diagram,

$$M_{tr} = (32\text{ kips})(20\text{ ft}) + (32\text{ kips} + 8\text{ kips})(13\text{ ft})$$

$$= 1160\text{ ft-kips [controls]}$$

$$M_{tandem} = (25\text{ kips})(20\text{ ft} + 18\text{ ft}) = 950\text{ ft-kips}$$

$$M_{ln} = \frac{\left(0.64\,\frac{\text{kips}}{\text{ft}}\right)(80\text{ ft})^2}{8} = 512\text{ ft-kips}$$

The maximum live load plus impact moment is defined by the following equation.

$$M_{LL+IM} = DFM_{mi}\left(\left(M_{tr}\text{ or }M_{tandem}\right)\left(1+\frac{IM}{100}\right)+M_{ln}\right)$$

$$= (0.716)((1160\text{ ft-kips})(1 + 0.33) + 512\text{ ft-kips})$$

$$= 1471.2\text{ ft-kips}$$

For the approximate maximum shear due to HL-93 loading, please see Figures 3.37 through 3.40, and use the influence line diagram.

$$V_{tr} = (32\text{ kips})(1 + 0.825) + (8\text{ kips})(0.65) = 63.60\text{ kips [controls]}$$

$$V_{tandem} = (25\text{ kips})(1 + 0.95) = 48.75\text{ kips}$$

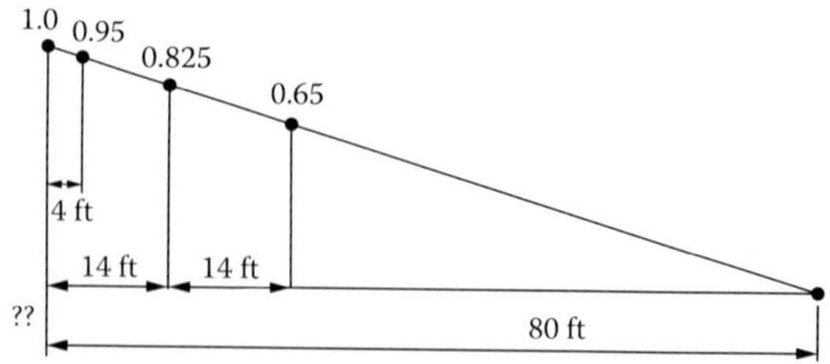

FIGURE 3.37
Influence line diagram for maximum shear at support.

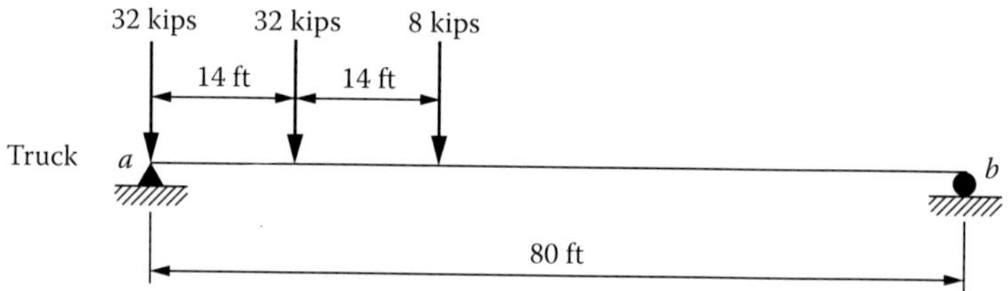

FIGURE 3.38
Design truck position for shear at support.

FIGURE 3.39
Design tandem load position for shear at support.

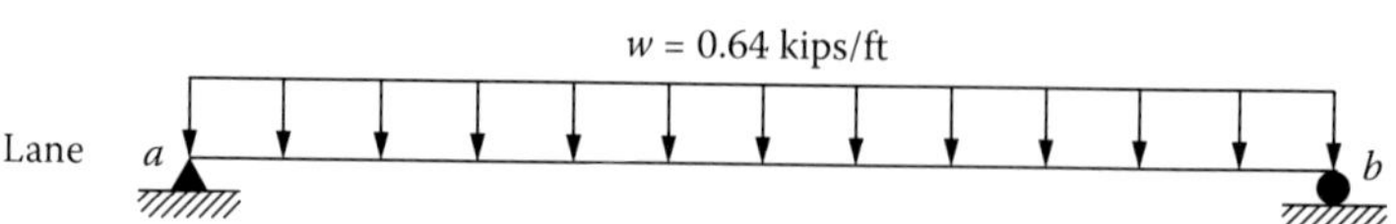

FIGURE 3.40
Design lane load for shear at support.

$$V_{ln} = \frac{\left(0.64\dfrac{\text{kips}}{\text{ft}}\right)(80\text{ ft})}{2} = 25.6\text{ kips}$$

The maximum live load plus impact shear is defined by the following equation:

$$V_{LL+IM} = DFV\left(\left(V_{tr} \text{ or } V_{tandem}\right)\left(1+\frac{IM}{100}\right)+V_{ln}\right)$$

$$= (0.814)((63.6 \text{ kips})(1 + 0.33) + 25.6 \text{ kips}) = 89.69 \text{ kips}$$

Interior Beam Summary (using the maximum dead and live/impact load, shear, and moment equations)

Note that $DC = DC_1 + DC_2$ or $w_{DC} = w_{DC1} + w_{DC2}$

$$V_{DC} = V_{max} = 40\ w_{DC} = (40)(1.769 \text{ kips/ft}) = 70.76 \text{ kips}$$

$$M_{DC} = M_{max} = 800\ w_{DC} = (800)(1.769 \text{ kips/ft}) = 1415.2 \text{ ft-kips}$$

Due to the slab and the beam weight,

$$V_{DC1} = 40\ w_{DC1} = (40)(1.60 \text{ kips/ft}) = 64.0 \text{ kips}$$

$$M_{DC1} = 800\ w_{DC1} = (800)(1.60 \text{ kips/ft}) = 1280 \text{ ft-kips}$$

Due to superimposed dead loads – parapet and curb loads,

$$V_{DC2} = 40\ w_{DC2} = (40)(0.169 \text{ kips/ft}) = 6.76 \text{ kips}$$

where:

DC_1 = beam weight + slab weight = 0.85 kips/ft + 0.75 kips/ft = 1.60 kips/ft

DC_2 = superimposed parapet and curb loads = 0.169 kips/ft

$M_{DC2} = 800\ w_{DC2} = (800)(0.169 \text{ kips/ft}) = 135.2 \text{ ft-kips}$

Due to the wearing surface weight,

$$V_{DW} = 40\ w_{DW} = (40)(0.240 \text{ kips/ft}) = 9.60 \text{ kips}$$

$$M_{DW} = 800\ w_{DW} = (800)(0.240 \text{ kips/ft}) = 192.0 \text{ ft-kips}$$

Please see Table 3.9.

Recall that:

w = beam weight = 0.85 kips/ft

w_D = slab weight = 0.75 kips/ft

$w_{DC1} = w + w_D$ = weights of beam and slab = 1.60 kips/ft

M_{DC1} = 1280.0 ft-kips

TABLE 3.9

Summary of Dead Load Moments and Shears

Load Type	w (kips/ft)	M (ft-kips)	V (kips)
DC	1.769	1,415.2	70.76
DC_1	1.60	1,280.0	64.0
DC_2	0.169	135.2	6.76
DW	0.240	192.0	9.60
LL + IM	N/A	1,471.2	89.69

V_{DC1} = 64.0 kips

w_s = w_{DC2} = superimposed dead loads (parapet and curb) = 0.169 kips/ft

M_{DC2} = 135.2 ft-kips

V_{DC2} = 6.76 kips

w_{DC} = w_{DC1} + w_{DC2} = weight of beam and slab plus parapet and curb = 1.769 kips/ft

M_{DC} = 1415.2 ft-kips

V_{DC} = 70.76 kips

w_{DW} = w_{FWS} = future wearing surface = 0.24 kips/ft

M_{DW} = 192.0 ft-kips

V_{DW} = 9.6 kips

S_{ncb} = section modulus of noncomposite section for the bottom extreme fiber

S_{nct} = section modulus of noncomposite section for the top extreme fiber

S_{cb} = section modulus of composite section for the bottom extreme fiber

S_{ct} = section modulus of composite section for the top extreme fiber

Step 5: Estimate the Required Prestress

Calculate the bottom tensile stress in prestressed concrete, f_{gb}, noncomposite, and composite properties.

Service III Limit State governs for longitudinal analysis relating to tension in prestressed concrete superstructures with the objective of crack control and to principal tension in the webs of segmental concrete girders.

A Art. 3.4.1

Service III Limit State

A Tbl. 3.4.1-1

$$Q = \eta((1.0)(DC + DW) + (0.8)(LL + IM))$$

$$\eta = 1.0$$

A Art. 1.3.2

$$f_{gb} = \frac{M_{DC1}}{S_{ncb}} + \frac{M_{DC2} + M_{DW} + 0.8(M_{LL+IM})}{S_{cb}}$$

$$= \frac{(1280.0 \text{ ft-kips})\left(\dfrac{12 \text{ in}}{\text{ft}}\right)}{10{,}083 \text{ in}^3}$$

$$+ \frac{135.2 \text{ ft-kips} + 192 \text{ ft-kips} + 0.8(1471.2 \text{ ft-kips})}{15{,}925 \text{ in}^3}$$

$$= 2.66 \text{ (2.7 ksi)}$$

Tensile stress limit at service limit state after losses, fully pretensioned is

A Art. 5.9.4.2.2; Tbl. 5.9.4.2.2-1

$$0.19\sqrt{f'_{cg}} = 0.19\sqrt{6.5 \text{ ksi}} = 0.484 \text{ ksi}$$

The excess tension in the bottom fiber due to applied loads is

$$f_t = f_{gb} - 0.19\sqrt{f'_{cg}} = 2.66 \text{ ksi} - 0.484 \text{ ksi} = 2.18 \text{ ksi}$$

The location of the center of gravity of the strands at midspan usually ranges from 5 to 15% of the beam depth. In this example, first assume 10%.

The distance between the center of gravity of the bottom strands to the bottom fiber is assumed to be

$$y_{bs} = (0.1)(54 \text{ in}) = 5.4 \text{ in}$$

The strand eccentricity at midspan is

$$e_c = y_b - y_{bs} = 25.31 \text{ in} - 5.4 \text{ in} = 19.91 \text{ in}$$

To find the initial required prestress force, P_i,

$$f_t = \frac{P_i}{A_g} + \frac{P_i e_c}{S_{ncb}}$$

$$2.18 \text{ ksi} = \frac{P_i}{816 \text{ in}^2} + \frac{P_i(19.91 \text{ in})}{10,083 \text{ in}^3}$$

$$P_i = 681.2 \text{ kips}$$

The stress in prestressing tendon immediately prior to transfer, f_{pbt}, is

A Art. 5.9.3

$$f_{pbt} = 0.75\ f_{pu} = (0.75)(270 \text{ ksi}) = 202.5 \text{ ksi}$$

A Tbl. 5.9.3-1

Assuming a 25% prestress loss, prestress force per ½ in strand after losses, f_{ps}, is

$$f_{ps} = A_{ps} f_{pbt}(1 - 0.25) = (0.153 \text{ in}^2)(202.5 \text{ ksi})(1 - 0.25) = 23.2 \text{ kips}$$

Number of strands required is

$$\frac{P_i}{f_{ps}} = \frac{681.2 \text{ kips}}{23.2 \text{ kips}} = 29.4$$

Try 30, ½ in strands.

Check the assumption of bottom strand center of gravity using the following configuration. Please see Figure 3.41.

The distance between the center of gravity of the 30 strands to the bottom fiber is

$$y_{bs} = \frac{(8)(2 \text{ in}) + (8)(4 \text{ in}) + (6)(6 \text{ in}) + (4)(8 \text{ in}) + (2)(10 \text{ in}) + (2)(12 \text{ in})}{30 \text{ strands}}$$

$$= 5.33 \text{ in}$$

Inasmuch as the assumed value for y_{bs} was 5.4 in, another iteration is not required.

The final strand eccentricity value is

$$e_c = y_b - y_{bs} = 25.31 \text{ in} - 5.33 \text{ in} = 19.98 \text{ in}$$

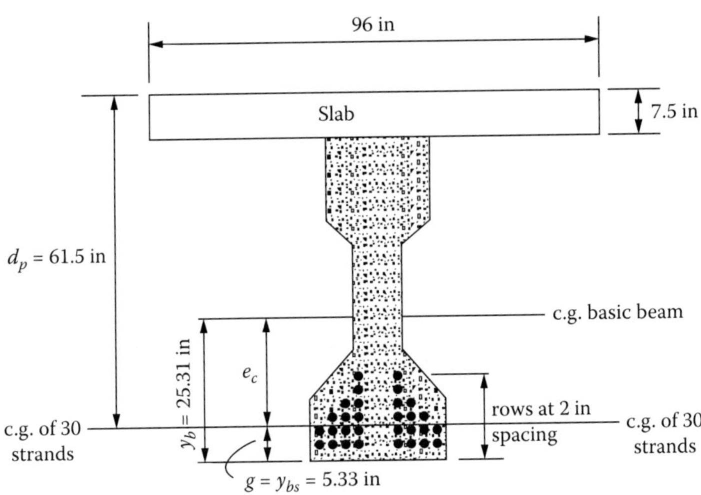

FIGURE 3.41
Prestressed I-beam with 30, ½ in strands.

Revised initial required prestress force using 30 strands as opposed to 29.4 strands.

$$\text{number of strands} = \frac{P_i}{f_{ps}}$$

$$30 \text{ strands} = \frac{P_i}{23.2 \text{ kips}}$$

$$P_i = 696 \text{ kips}$$

Area of prestressing steel is

$$(30 \text{ strands})(0.153 \text{ in}^2) = 4.59 \text{ in}^2$$

Step 6: Calculate the Prestress Losses

In pretensioned members, $\Delta f_{pT} = \Delta f_{pES} + \Delta f_{pLT}$

A Eq. 5.9.5.1-1

- Δf_{pT} total loss kips/in²
- Δf_{pES} sum of all losses or gains due to elastic shortening or extension at the time of application of prestress and/or external loads
- Δf_{pLT} losses due to long-term shrinkage and creep of concrete, and relaxation of the steel kips/in²

Temporary allowable concrete stresses before losses due to creep and shrinkage are as follows (i.e., at time of initial prestress).

Find the loss due to elastic shortening, Δf_{pES}.

A Eq. 5.9.5.2.3a-1

$$\Delta f_{pES} = \frac{E_p}{E_{ci}} f_{cgp}$$

f_{cgp} is the concrete stress at the center of gravity of prestressing tendons due to prestressing force immediately after transfer and self-weight of member at the section of maximum moment.

f_{cgp} is found using an iterative process. Alternatively, the following calculation can be used to find Δf_{pES} for loss in prestressing steel due to elastic shortening.

A Eq. C5.9.5.2.3a-1

$$\Delta f_{pES} = \frac{A_{ps} f_{pbt}\left(I_g + e_m^2 A_g\right) - e_m M_g A_g}{A_{ps}\left(I_g + e_m^2 A_g\right) + \dfrac{A_g I_g E_{ci}}{E_p}}$$

A_{ps}	area of prestressing steel	4.59 in^2
A_g	gross area of section	816 in^2
E_{ci}	modulus of elasticity of concrete at transfer	4496 ksi
E_p	modulus of elasticity of prestressing tendons	28,500 ksi
$e_m = e_c$	average prestressing steel eccentricity at midspan	19.98 in
I_g	moment of inertia of the basic beam section	255,194 in^4

The midspan moment due to beam self weight is

$$M_g = \frac{wL^2}{8} = \frac{\left(0.85 \frac{\text{kips}}{\text{ft}}\right)(80\ \text{ft})^2}{8}$$

$$= 680\ \text{ft-kips}$$

The loss due to elastic shortening is

A Eq. C5.9.5.2.3a-1

$$\Delta f_{pES} = \frac{A_{ps}f_{pbt}\left(I_g + e_m^2 A_g\right) - e_m M_g A_g}{A_{ps}\left(I_g + e_m^2 A_g\right) + \dfrac{A_g I_g E_{ci}}{E_p}}$$

$$= \frac{\left(4.59\text{ in}^2\right)\left(202.5\text{ ksi}\right)\left(255{,}194\text{ in}^4 + \left(19.98\text{ in}\right)^2\left(816\text{ in}^2\right)\right) - \left(19.98\text{ in}\right)\left(680\text{ ft-kips}\right)\left(\dfrac{12\text{ in}}{1\text{ ft}}\right)\left(816\text{ in}^2\right)}{\left(4.59\text{ in}^2\right)\left(255{,}194\text{ in}^4 + \left(19.98\text{ in}\right)^2\left(816\text{ in}^2\right)\right) + \dfrac{\left(816\text{ in}^2\right)\left(255{,}194\text{ in}^4\right)\left(4496\text{ ksi}\right)}{28{,}500\text{ ksi}}}$$

$$= 11.46\text{ ksi}$$

Calculate the long-term prestress loss due to creep of concrete shrinkage of concrete, and relaxation of steel, Δf_{pLT}.

A Art. 5.9.5.3; Eq. 5.9.5.3-1

Where:

f_{pi} = f_{pbt} = stress in prestressing steel immediately prior to transfer
Δf_{pR} = relaxation loss

Δf_{pR} is an estimate of relaxation loss taken as 2.4 ksi for low relaxation strand, 10.0 ksi for stress relieved strand, and in accordance with manufacturer's recommendation for other types of strand.

A Art. 5.9.5.3

γ_h = correction factor for relative humidity

$\gamma_h = 1.7 - 0.01\text{ H} = 1.7 - (0.01)(0.72) = 1.69$

A Eq. 5.9.5.3-2

The average annual ambient relative humidity, H, in western Pennsylvania is approximately 72%.

A Fig. 5.4.2.3.3-1

The correction factor for specified concrete strength at the time of the prestress transfer to the concrete is,

A Eq. 5.9.5.3-3

$$\gamma_{st} = \frac{5}{(1+f'_{ci})} = \frac{5}{(1+5.5)}$$

$$= 0.77$$

$$\Delta f_{pLT} = (10.0)\frac{f_{pi}A_{ps}}{A_g}\gamma_h\gamma_{st} + (12.0)\gamma_h\gamma_{st} + \Delta f_{pR}$$

A Eq. 5.9.5.3-1

$$= (10.0)\left(\frac{(202.5\text{ ksi})(4.59\text{ in}^2)}{816\text{ in}^2}\right)(1.69)(0.77) + (12)(1.69)(0.77) + 2.4\text{ ksi}$$

$$= 32.84\text{ ksi}$$

The total prestress losses are

$$\Delta f_{pT} = \Delta f_{pES} + \Delta f_{pLT} = 11.46\text{ ksi} + 32.84\text{ ksi} = 44.30\text{ ksi}$$

A Eq. 5.9.5.1-1

Summary of Prestress Forces

The prestress stress per strand before transfer f_{pbt} is

A Tbl. 5.9.3-1

$$f_{pbt} = 0.75\ f_{pu}$$

$$= (0.75)\left(270\frac{\text{kips}}{\text{in}^2}\right) = 202.5\frac{\text{kips}}{\text{in}^2}$$

The prestress force per strand before transfer is

$$P_{pi} = (f_{pbt})A_{ps}$$

$$= (202.5\text{ ksi})(0.153\text{ in}^2) = 31\text{ kips}$$

The prestress force per strand after all losses is

$$P_{pe} = (202.5\text{ ksi} - 44.30\text{ ksi})(0.153\text{ in}^2) = 24.20\text{ kips}$$

Check assumption of 25% prestress losses.

$$\%\text{loss} = 1.0 - \frac{\text{prestress force after all losses}(= P_{pe})}{\text{prestress force before transfer}(= P_{pi})} = 1 - \frac{24.20 \text{ kips}}{31 \text{ kips}} = 21.9\%$$

Assumption of 25% was conservative because 21.9% < 25% [OK].

Step 7: Check Tensile and Compressive Concrete Stresses at Transfer

The compressive stress of concrete at time of prestressing before losses is

A Art. 5.9.4.1.1

$$f_{ci} = 0.60\, f'_{ci} = (0.6)(5.5 \text{ ksi}) = 3.3 \text{ ksi}$$

The tensile stress in prestressed concrete before losses is

A Art. 5.9.4.1.2; Tbl. 5.9.4.1.2-1

$$f_{ti} = 0.24\sqrt{f'_{ci}} = 0.24\sqrt{5.5 \text{ ksi}} = 0.563 \text{ ksi}$$

Stress at bottom of girder (compressive stress)

$$f_{bot} = -\frac{P_i}{A_g} - \frac{P_i e_c}{S_{ncb}} + \frac{M_g}{S_{ncb}}$$

$$= -\frac{696 \text{ kips}}{816 \text{ in}^2} - \frac{(696 \text{ kips})(19.98 \text{ in})}{10{,}083 \text{ in}^3} + \frac{680 \text{ ft-kips}}{10{,}083 \text{ in}^3}\left(\frac{12 \text{ in}}{\text{ft}}\right)$$

$$= -1.42 \text{ ksi} < -3.3 \text{ ksi [OK]}$$

Stress at top of girder (tensile stress)

$$f_{bot} = -\frac{P_i}{A_g} + \frac{P_i e_c}{S_{nct}} - \frac{M_g}{S_{nct}}$$

$$= -\frac{696 \text{ kips}}{816 \text{ in}^2} + \frac{(696 \text{ kips})(19.98 \text{ in})}{8895.0 \text{ in}^3} - \frac{680 \text{ ft-kips}}{8895.0 \text{ in}^3}\left(\frac{12 \text{ in}}{\text{ft}}\right)$$

$$= -0.207 \text{ ksi} < -0.563 \text{ ksi [OK]}$$

Step 8: Determine the Flexural Resistance Using the Strength I Limit State

Strength I Limit State, U = 1.25 DC + 1.50 DW + 1.75(LL + IM).

A Tbl. 3.4.1-1

Recall that DC is the weight due to the deck slab, the basic beam, and the parapet curb, and that DW is the weight due to the future wearing surface.

For these values, the Strength I Limit State can be written as follows:

$$M_u = (1.25)(1415.2 \text{ ft-kips}) + (1.50)(192 \text{ ft-kips}) + (1.75)(1471.2 \text{ ft-kips})$$

$$= 4631.6 \text{ ft-kips}$$

Find the average stress in prestressing steel assuming rectangular behavior.

A Eq. 5.7.3.1.1-1

$$f_{ps} = f_{pu}\left(1 - k\frac{c}{d_p}\right)$$

where:

d_p = distance from extreme compression fiber to the centroid of the prestressing tendons

f_{pu} = 270 ksi

k = 0.38

A Tbl. C5.7.3.1.1-1

$d_p = (h - y_{bs}\ (= g) + t_s)$
$= (54 \text{ in} - 5.33 \text{ in}) + 7.5 \text{ in}$
$= 56.17 \text{ in}$

$$c = \frac{(A_{ps})(f_{pu}) + A_s f_s - A'_s f'_s}{0.85\, f'_c\, \beta_1 b + k(A_{ps})\left(\frac{f_{pu}}{d_p}\right)}$$

A Eq. 5.7.3.1.1-4

where:

$$\beta_1 = 0.85 - 0.05(f'_{cg} - 4 \text{ ksi})$$

A Art. 5.7.2.2

$$= 0.85 - 0.05(6.5 \text{ ksi} - 4 \text{ ksi})$$

$$= 0.725$$

$$c = \frac{(4.59\ \text{in}^2)(270\ \text{ksi})}{0.85(6.5\ \text{ksi})(0.725)(96\ \text{in}) + (0.38)(4.59\ \text{in}^2)\left(\dfrac{270\ \text{ksi}}{56.17\ \text{in}}\right)}$$

$$= 3.16\ \text{in} < t_s = 7.5\ \text{in}$$

so the assumption is OK.

The average stress in prestressing steel when the nominal resistance of member is required, f_{ps},

$$f_{ps} = (270\ \text{ksi})\left(1 - (0.38)\frac{3.16\ \text{in}}{56.17\ \text{in}}\right) = 264.2\ \text{ksi}$$

Find the factored flexural resistance for flanged section:

A Art. 5.7.3.2.2

$$a = \beta_1 c = (0.725)(3.16\ \text{in})$$

$$= 2.29\ \text{in}$$

The factored resistance M_r shall be taken as:

A Eq. 5.7.3.2.2

$$M_r = \Phi M_n$$

where:

M_n = nominal resistance

Φ = resistance factor = 1.0

A Art. 5.5.4.2

$M_r = (1.0)(M_n) = M_n$

A Eq. 5.7.3.2.2-1

$$M_n = A_{ps} f_{ps}\left(d_p - \frac{a}{2}\right)$$

$$M_r = M_n$$

$$M_r = (4.59\ \text{in}^2)(264.2\ \text{ksi})\left(56.17\ \text{in} - \frac{2.29\ \text{in}}{2}\right)\left(\frac{1\ \text{ft}}{12\ \text{in}}\right)$$

$$M_r = 5560\ \text{ft-kips} > M_u = 4631.6\ \text{ft-kips}\ [\text{OK}]$$

Appendix A: Distribution of Live Loads per Lane for Moment in Interior Beams (AASHTO Table 4.6.2.2.2b-1)

Distribution of Live Loads per Lane for Moment in Interior Beams

Type of Superstructure	Applicable Cross Section from Table 4.6.2.2.1-1	Distribution Factors	Range of Applicability
Wood Deck on Wood or Steel Beams	a, 1	See **Table 4.6.2.2.2a-1**	
Concrete Deck on Wood Beams	1	One Design Lane Loaded: $S/12.0$ Two or More Design Lanes Loaded: $S/10.0$	$S \le 6.0$
Concrete Deck, Filled Grid, Partially Filled Grid, or Unfilled Grid Deck Composite with Reinforced Concrete Slab on Steel or Concrete Beams; Concrete T-Beams, T- and Double T-Sections	a, e, k, and also i, j if sufficiently connected to act as a unit	One Design Lane Loaded: $0.06+\left(\frac{S}{14}\right)^{0.4}\left(\frac{S}{L}\right)^{0.3}\left(\frac{K_g}{12.0\,Lt_s^3}\right)^{0.1}$ Two or More Design Lanes Loaded: $0.075+\left(\frac{S}{9.5}\right)^{0.6}\left(\frac{S}{L}\right)^{0.2}\left(\frac{K_g}{12.0\,Lt_s^3}\right)^{0.1}$	$3.5 \le S \le 16.0$ $4.5 \le t_s \le 12.0$ $20 \le L \le 240$ $N_b \ge 4$ $10{,}000 \le K_g \le 7{,}000{,}000$
		Use lesser of the values obtained from the equation above with $N_b = 3$ or the lever rule	$N_b = 3$
Cast-in-Place Concrete Multicell Box	d	One Design Lane Loaded: $\left(1.75+\frac{S}{3.6}\right)\left(\frac{1}{L}\right)^{0.35}\left(\frac{1}{N_c}\right)^{0.45}$ Two or More Design Lanes Loaded: $\left(\frac{13}{N_c}\right)^{0.3}\left(\frac{S}{5.8}\right)\left(\frac{1}{L}\right)^{0.25}$	$7.0 \le S \le 13.0$ $60 \le L \le 240$ $N_c \ge 3$ If $N_c > 8$ use $N_c = 8$

Concrete Deck on Concrete Spread Box Beams	b, c	One Design Lane Loaded: $\left(\frac{S}{3.0}\right)^{0.35}\left(\frac{Sd}{12.0\,L^2}\right)^{0.25}$ Two or More Design Lanes Loaded: $\left(\frac{S}{6.3}\right)^{0.6}\left(\frac{Sd}{12.0\,L^2}\right)^{0.125}$	$6.0 \le S \le 18.0$ $20 \le L \le 140$ $18 \le d \le 65$ $N_b \ge 3$
		Use Lever Rule	$S > 18.0$
Concrete Beams used in Multibeam Decks	f	One Design Lane Loaded: $k\left(\frac{b}{33.3\,L}\right)^{0.5}\left(\frac{I}{J}\right)^{0.25}$ where: $k = 2.5(N_b)^{-0.2} \ge 1.5$ Two or More Design Lanes Loaded: $k\left(\frac{b}{305}\right)^{0.6}\left(\frac{b}{12.0\,L}\right)^{0.2}\left(\frac{I}{J}\right)^{0.06}$	$35 \le b \le 60$ $20 \le L \le 120$ $5 \le N_b \le 20$
	g if sufficiently connected to act as a unit		

Source: **AASHTO Table 4.6.2.2.2b-1.**

Appendix B: Distribution of Live Loads per Lane for Moment in Exterior Longitudinal Beams (AASHTO *Table 4.6.2.2.2d-1)*

Distribution of Live Loads per Lane for Moment in Exterior Longitudinal Beams

Type of Superstructure	Applicable Cross Section from **Table 4.6.2.2.1-1**	One Design Lane Loaded	Two or More Design Lanes Loaded	Range of Applicability
Wood Deck on Wood or Steel Beams	a, 1	Lever Rule	Lever Rule	N/A
Concrete Deck on Wood Beams	1	Lever Rule	Lever Rule	N/A
Concrete Deck, Filled Grid, Partially Filled Grid, or Unfilled Grid Deck Composite with Reinforced Concrete Slab on Steel or Concrete Beams; Concrete T-Beams, T- and Double T-Sections	a, e, k, and also i, j if sufficiently connected to act as a unit	Lever Rule	$g = e\, g_{interior}$ $e = 0.77 + \frac{d_e}{9.1}$	$-1.0 \le d_e \le 5.5$
			Use lesser of the values obtained from the equation above with $N_b = 3$ or the lever rule	$N_b = 3$
Cast-in-Place Concrete Multicell Box	d	$g = \frac{W_e}{14}$	$g = \frac{W_e}{14}$	$W_e \le S$
		or the provisions for a whole-width design specified in **Article 4.6.2.2.1**		
Concrete Deck on Concrete Spread Box Beams	b, c	Lever Rule	$g = e\, g_{interior}$ $e = 0.97 + \frac{d_e}{28.5}$	$0 \le d_e \le 4.5$ $6.0 < S \le 18.0$
			Use Lever Rule	$S > 18.0$

Concrete Box Beams Used in Multibeam Decks	f, g	$g = e\ g_{interior}$ $e = 1.125 + \frac{d_e}{30} \geq 1.0$	$g = e\ g_{interior}$ $e = 1.04 + \frac{d_e}{25} \geq 1.0$	$d_e \leq 2.0$
Concrete Beams Other than Box Beams Used in Multibeam Decks	h	Lever Rule	Lever Rule	N/A
	i, j if connected only enough to prevent relative vertical displacement at the interface			
Open Steel Grid Deck on Steel Beams	a	Lever Rule	Lever Rule	N/A
Concrete Deck on Multiple Steel Box Girders	b, c	As specified in **Table 4.6.2.2.2b-1**		

Source: **AASHTO Table 4.6.2.2.2d-1.**

Appendix C: Distribution of Live Load per Lane for Shear in Interior Beams (AASHTO Table 4.6.2.2.3a-1)

Distribution of Live Load per Lane for Shear in Interior Beams				
Type of Superstructure	**Applicable Cross Section from Table 4.6.2.2.1-1**	**One Design Lane Loaded**	**Two or More Design Lanes Loaded**	**Range of Applicability**
Wood Deck on Wood or Steel Beams	a, 1	See **Table 4.6.2.2.2a-1**		
Concrete Deck on Wood Beams	1	Lever Rule	Lever Rule	N/A
Concrete Deck, Filled Grid, Partially Filled Grid, or Unfilled Grid Deck Composite with Reinforced Concrete Slab on Steel or Concrete Beams; Concrete T-Beams, T- and Double T-Sections	a, e, k, and also i, j if sufficiently connected to act as a unit	$0.36+\frac{S}{25.0}$	$0.2+\frac{S}{12}\left(\frac{S}{35}\right)^{2.0}$	$3.5 \le S \le 16.0$ $20 \le L \le 240$ $4.5 \le t_s \le 12.0$ $N_b \ge 4$
		Lever Rule	Lever Rule	$N_b = 3$
Cast-in-Place Concrete Multicell Box	d	$\left(\frac{S}{9.5}\right)^{0.6}\left(\frac{d}{12.0\,L}\right)^{0.1}$	$\left(\frac{S}{7.3}\right)^{0.9}\left(\frac{d}{12.0\,L}\right)^{0.1}$	$6.0 \le S \le 13.0$ $20 \le L \le 240$ $35 \le d \le 110$ $N_c \ge 3$
Concrete Deck on Concrete Spread Box Beams	b, c	$\left(\frac{S}{10}\right)^{0.6}\left(\frac{d}{12.0\,L}\right)^{0.1}$	$\left(\frac{S}{7.4}\right)^{0.8}\left(\frac{d}{12.0\,L}\right)^{0.1}$	$6.0 \le S \le 18.0$ $20 \le L \le 140$ $18 \le d \le 65$ $N_b \ge 3$
		Lever Rule	Lever Rule	$S > 18.0$

Concrete Box Beams Used in Multibeam Decks	f, g	$\left(\frac{b}{130\,L}\right)^{0.15}\left(\frac{I}{J}\right)^{0.05}$	$\left(\frac{b}{156}\right)^{0.4}\left(\frac{b}{12.0\,L}\right)^{0.1}\left(\frac{I}{J}\right)^{0.05}\left(\frac{b}{48}\right)$ $\frac{b}{48} \geq 1.0$	$35 \leq b \leq 60$ $20 \leq L \leq 120$ $5 \leq N_b \leq 20$ $25{,}000 \leq J \leq 610{,}000$ $40{,}000 \leq I \leq 610{,}000$
Concrete Beams Other Than Box Beams Used in Multibeam Decks	h			
	i, j if connected only enough to prevent relative vertical displacement at the interface	Lever Rule	Lever Rule	N/A
Open Steel Grid Deck on Steel Beams	a	Lever Rule	Lever Rule	N/A
Concrete Deck on Multiple Steel Box Beams	b, c	As specified in **Table 4.6.2.2.2b-1**		

Source: **AASHTO Table 4.6.2.2.3a-1.**

Appendix D: Distribution of Live Load per Lane for Shear in Exterior Beams (AASHTO Table 4.6.2.2.3b-1)

Distribution of Live Load per Lane for Shear in Exterior Beams

Type of Superstructure	Applicable Cross Section from Table 4.6.2.2.1-1	One Design Lane Loaded	Two or More Design Lanes Loaded	Range of Applicability
Wood Deck on Wood or Steel Beams	a, 1	Lever Rule	Lever Rule	N/A
Concrete Deck on Wood Beams	1	Lever Rule	Lever Rule	N/A
Concrete Deck, Filled Grid, Partially Filled Grid, or Unfilled Grid Deck Composite with Reinforced Concrete Slab on Steel or Concrete Beams; Concrete T-Beams, T- and Double T-Sections	a, e, k, and also i, j if sufficiently connected to act as a unit	Lever Rule	$g = e\ g_{interior}$ $e = 0.6 + \frac{d_e}{10}$	$-1.0 \le d_e \le 5.5$
			Lever Rule	$N_b = 3$
Cast-in-Place Concrete Multicell Box	d	Lever Rule	$g = e\ g_{interior}$ $e = 0.64 + \frac{d_e}{12.5}$	$-2.0 \le d_e \le 5.0$
		or the provisions for a whole-width design specified in **Article 4.6.2.2.1**		
Concrete Deck on Concrete Spread Box Beams	b, c	Lever Rule	$g = e\ g_{interior}$ $e = 0.8 + \frac{d_e}{10}$	$0 \le d_e \le 4.5$
			Lever Rule	$S > 18.0$

Concrete Box Beams Used in Multibeam Decks	f, g	$g = e\, g_{interior}$ $e = 1.25 + \frac{d_e}{20} \geq 1.0$	$g = e g_{interior}\left(\frac{48}{b}\right)$ $\frac{48}{b} \leq 1.0$ $e = 1 + \left(\frac{d_e + \frac{b}{12} - 2.0}{40}\right)^{0.5} \geq 1.0$	$d_e \leq 2.0$ $35 \leq b \leq 60$
Concrete Beams Other than Box Beams Used in Multibeam Decks	h i, j if connected only enough to prevent relative vertical displacement at the interface	Lever Rule	Lever Rule	N/A
Open Steel Grid Deck on Steel Beams	a	Lever Rule	Lever Rule	N/A
Concrete Deck on Multiple Steel Box Beams	b, c	As specified in **Table 4.6.2.2.2b-1**		
Source: **AASHTO Table 4.6.2.2.3b-1.**				

Appendix E: U.S. Customary Units and Their SI Equivalents

U.S. Customary Units and Their SI Equivalents		
Quantity	**U.S. Customary Units**	**SI Equivalent**
Area	ft^2	0.0929 m^2
	in^2	645.2 mm^2
Force	kip	4.448 kN
	lbf	4.448 N
Length	ft	0.3048 m
	in	25.40 mm
Moment	ft-lbf	1.356 N-m
	ft-kip	1.355×10^{4} kN-m
	in-lbf	0.1130 N-m
	in-kip	1.130×10^4 kN-m
Moment of Inertia	in^4	0.4162×10^6 mm^4
Stress	lbf/ft^2	47.88 Pa
	lbf/in^2 (psi)	6.895 kPa
	kip/in^2 (ksi)	6895.0 kPa

References

Primary References

The following sources were used to create this book and can serve as primary references for working bridge design-related problems for the NCEES civil and structural PE exams. (If a reference is referred to in this book by a name other than its title, the alternative name is indicated in brackets following the reference.)

American Association of State Highway and Transportation Officials. *AASHTO LRFD Bridge Design Specifications*, 5th ed., Washington, DC
2010. [AASHTO]

American Association of State Highway and Transportation Officials. *Manual for Bridge Evaluation.* 2nd ed. Washington, DC: 2011. [MBE-2]

American Association of State Highway and Transportation Officials. *AASHTO LRFD Bridge Design Specifications*, 4th ed. Washington, DC: 2007 (with 2008 Interim Revisions). [AASHTO]

American Association of State Highway and Transportation Officials. *Standard Specifications for Highway Bridges.* 17th ed. Washington, DC: 2002.

American Institute of Steel Construction. *Steel Construction Manual.* 13th ed. Chicago: 2005. [AISC Manual]

Supplementary References

The following additional references were used to create this book.

Barker, Richard M., and Jay A. Pucket. *Design of Highway Bridges, an LRFD Approach*, 2nd edition, Hoboken: John Wiley & Sons, Inc. 2007.

Kim, Robert H., and Jai B. Kim. *Bridge Design for the Civil and Structural Professional Engineering Exams*, 2nd ed., Belmont, CA: Professional Publications, Inc. 2001.

Kim, Robert H., and Jai B. Kim. Relocation of Coudersport 7th Street Bridge (historic truss) to 4th Street over Allegheny River, Potter County, Pennsylvania. Unpublished report for Pennsylvania State Department of Transportation. 2007–2009.

Schneider, E.F., and Shrinivas B. Bhide. Design of concrete bridges by the AASHTO LRFD specifications and LRFD design of cast-in-place concrete bridges. Report for the Portland Cement Association. Skokie, IL: July 31–August 3, 2007.

Sivakumar, B. Design and evaluation of highway bridge superstructures using LRFD. Unpublished American Society of Civil Engineers seminar notes. Phoenix, AZ: 2008.

Index

F

K

L

M

U

V

W

NIVERSITIES AT MEDWAY